T0296514

LONDON MATHEMATICAL SOCIETY STUDENT TEXTS

3 Local fields, J.W.S. CASSELS
4 An introduction to twistor theory: Second edition, S.A. HUGGETT & K.P. TOD
5 Introduction to general relativity, L.P. HUGHSTON & K.P. TOD
7 The theory of evolution and dynamical systems, J. HOFBAUER & K. SIGMUND
8 Summing and nuclear norms in Banach space theory, G.J.O. JAMESON
9 Automorphisms of surfaces after Nielsen and Thurston, A. CASSON & S. BLEILER
11 Spacetime and singularities, G. NABER
12 Undergraduate algebraic geometry, MILES REID
13 An introduction to Hankel operators, J.R. PARTINGTON
15 Presentations of groups: Second edition, D.L. JOHNSON
17 Aspects of quantum field theory in curved spacetime, S.A. FULLING
18 Braids and coverings: selected topics, VAGN LUNDSGAARD HANSEN
19 Steps in commutative algebra, R.Y. SHARP
20 Communication theory, C.M. GOLDIE & R.G.E. PINCH
21 Representations of finite groups of Lie type, FRANÇOIS DIGNE & JEAN MICHEL
22 Designs, graphs, codes, and their links, P.J. CAMERON & J.H. VAN LINT
23 Complex algebraic curves, FRANCES KIRWAN
24 Lectures on elliptic curves, J.W.S. CASSELS
25 Hyperbolic geometry, BIRGER IVERSEN
26 An introduction to the theory of L-functions and Eisenstein series, H. HIDA
27 Hilbert Space: compact operators and the trace theorem, J.R. RETHERFORD
28 Potential theory in the complex plane, T. RANSFORD
29 Undergraduate commutative algebra, M. REID
31 The Laplacian on a Riemannian manifold, S. ROSENBERG
32 Lectures on Lie groups and Lie algebras, R. CARTER, G. SEGAL & I. MACDONALD
33 A primer of algebraic *D*-modules, S.C. COUTINHO
34 Complex algebraic surfaces, A. BEAUVILLE
35 Young tableaux, W. FULTON
37 A mathematical introduction to wavelets, P. WOJTASZCZYK
38 Harmonic maps, loop groups and integrable systems, M. GUEST
39 Set theory for the working mathematician, K. CIESIELSKI
40 Ergodic theory and dynamical systems, M. POLLICOTT & M. YURI
41 The algorithmic resolution of diophantine equations, N.P. SMART
42 Equilibrium states in ergodic theory, G. KELLER

London Mathematical Society Student Texts 45

Permutation Groups

Peter J. Cameron
Queen Mary and Westfield College, London

CAMBRIDGE UNIVERSITY PRESS
Cambridge, New York, Melbourne, Madrid, Cape Town, Singapore, São Paulo

Cambridge University Press
The Edinburgh Building, Cambridge CB2 2RU, UK

Published in the United States of America by Cambridge University Press, New York

www.cambridge.org
Information on this title: www.cambridge.org/9780521653022

First published 1999

A catalogue record for this publication is available from the British Library

ISBN-13 978-0-521-65302-2 hardback
ISBN-10 0-521-65302-9 hardback

ISBN-13 978-0-521-65378-7 paperback
ISBN-10 0-521-65378-9 paperback

Transferred to digital printing 2005

Contents

Preface **ix**

1 General theory **1**
1.1 History . . . 1
1.2 Actions and G-spaces . . . 2
1.3 Orbits and transitivity . . . 3
1.4 Transitive groups and coset spaces . . . 4
1.5 Sylow's Theorem . . . 6
1.6 Regular groups . . . 7
1.7 Groups with regular normal subgroups . . . 9
1.8 Multiple transitivity . . . 10
1.9 Primitivity . . . 11
1.10 Wreath products . . . 11
1.11 Orbitals . . . 13
1.12 Sharp k-transitivity . . . 15
1.13 The Schreier–Sims algorithm . . . 17
1.14 Jerrum's filter . . . 19
1.15 The length of S_n . . . 20
1.16 At the keyboard . . . 21
1.17 Appendix: Cycles and parity . . . 25
1.18 Exercises . . . 27

2 Representation theory **35**
2.1 Historical note . . . 35
2.2 Centraliser algebra . . . 36
2.3 The Orbit-Counting Lemma . . . 37
2.4 Character theory . . . 40

2.5 The permutation character 42
2.6 The diagonal group 44
2.7 Frobenius–Schur index 45
2.8 Parker's Lemma 47
2.9 Characters of abelian groups 51
2.10 Characters of the symmetric group 53
2.11 Appendix: Möbius inversion 56
2.12 Exercises 57

3 Coherent configurations **63**
3.1 Introduction 63
3.2 Algebraic theory 66
3.3 Association schemes 68
3.4 Algebra of association schemes 73
3.5 Example: Strongly regular graphs 76
3.6 The Hoffman–Singleton graph 79
3.7 Automorphisms 83
3.8 Valency bounds 85
3.9 Distance-transitive graphs 87
3.10 Multiplicity bounds 90
3.11 Duality 92
3.12 Wielandt's Theorem 94
3.13 Exercises 95

4 The O'Nan–Scott Theorem **99**
4.1 Introduction 99
4.2 Precursors 100
4.3 Product action and basic groups 102
4.4 Some basic groups 104
4.5 The O'Nan–Scott Theorem 105
4.6 Maximal subgroups of S_n 107
4.7 The finite simple groups 108
4.8 Application: Multiply-transitive groups 110
4.9 Application: Degrees of primitive groups 111
4.10 Application: Orders of primitive groups 112
4.11 Application: The length of S_n 117
4.12 Application: Distance-transitive graphs 118
4.13 Bases 120
4.14 Geometric groups and IBIS groups 123
4.15 Appendix: Matroids 125
4.16 Exercises 127

5 Oligomorphic groups **131**
5.1 The random graph 131
5.2 Oligomorphic groups 135
5.3 First-order logic 135
5.4 Automorphism groups and topology 138
5.5 Countably categorical structures 139
5.6 Homogeneous structures 141
5.7 Cycle index 143
5.8 A graded algebra 145
5.9 Monotonicity 147
5.10 Set-transitive groups 148
5.11 Growth rates 150
5.12 On complementation and switching 153
5.13 Appendix: Cycle index 157
5.14 Exercises 160

6 Miscellanea **165**
6.1 Finitary permutation groups 165
6.2 Neumann's Lemma 167
6.3 Cofinitary permutation groups 168
6.4 Theorems of Blichfeldt and Maillet 170
6.5 Cycle-closed permutation groups 173
6.6 Fixed-point-free elements 173
6.7 The Orbit-Counting Lemma revisited 176
6.8 Jordan groups 179
6.9 Orbits on moieties 182
6.10 Exercises 183

7 Tables **187**
7.1 Simple groups of Lie type 188
7.2 Sporadic simple groups 192
7.3 Affine 2-transitive groups 194
7.4 Almost simple 2-transitive groups 196
7.5 Exercises 198

Bibliography **199**

Index **213**

Preface

This book grew from a course I gave at the Euler Institute for Discrete Mathematics and its Applications in the week 10–14 November 1997. The audience were keen to have notes, and so handwritten notes were produced each day. I have corrected and expanded these notes into the present version. The comments, questions and perplexities of the students on the course have been of very great value to me. In particular, they wanted to see proofs of theorems wherever possible, so I have included more proofs than I did in the lectures. Where a proof is too long to explain completely, I have given a sketch which tries to convey the main ideas, or worked a special case. Five chapters correspond to the five days of the course: introduction, character theory, the O'Nan–Scott Theorem, oligomorphic groups, and miscellanea.

Since the course, I have added a chapter on coherent configurations, two sections on computations with permutation groups, and tables of the finite simple groups and the finite 2-transitive groups.

I assume some knowledge of group theory and some mathematical sophistication. When other areas of mathematics (such as probability or logic) are invoked, I have tried not to assume any detailed knowledge: some results can be taken on trust.

The choice of topics is a bit idiosyncratic; this is not a complete treatment of the subject. Other topics which might have been chosen, and further details of those which do appear, are given in the references. For about thirty years, the only general reference on permutation groups was Helmut Wielandt's influential book [186]. In addition, Wielandt wrote several sets of lecture notes and many important papers; these are conveniently available now in Volume 1 of his *Mathematische Werke* [189]. In the last few years, more sources have become available, and I recommend the following as general references: the books by Bhattacharjee, Macpherson, Möller and

Neumann [17], Cameron [34], and Dixon and Mortimer [64], and the chapter on Permutation Groups in the Handbook of Combinatorics [36]. Passman's book [146] discusses such topics as the detailed structure of Frobenius and Zassenhaus groups.

I am grateful to EIDMA (especially Henny Houben and Henk van Tilborg) for the opportunity to give the course; to the students, especially Jürgen Müller and Max Neunhöffer, whose construction of the random graph is included in Chapter 5; to Colva Roney-Dougal, for working through the text and exercises, spotting many misprints and inclarities; to Leonard Soicher, for help with the worked examples using GAP; to Sasha Ivanov and Joachim Neubüser, for the history of cellular algebras and of classification of permutation groups; and to two anonymous referees, for helpful comments.

I have provided many exercises. Some of them are not straightforward; in cases where these are published results, references to the literature have been given. Hints are usually included.

In several cases, I have referred to sources of information available on the Internet. There is also a World Wide Web page associated with the book, at the URL

`http://www.maths.qmw.ac.uk/~pjc/permgps/`

This will contain the links mentioned in the book (and others), and possibly solutions to the exercises.

The GAP programming examples in Chapters 1 and 3 have been tested with the current release (version 3.4.4) of GAP. They also work with the beta4 version of GAP 4 (using an unofficial version of GRAPE). It is hoped that, by the time this book is published, GAP 4 will be officially available and the examples will work correctly. News of this will be posted on the Web page.

Peter J. Cameron
9 October 1998

CHAPTER 1

General theory

1.1 History

Until about 1850, the term 'group' referred to a set G of transformations of a set Ω, such that G is closed under composition, contains the identity transformation, and contains the inverse of each of its elements. This is what would now be called a 'permutation group'. (Because of the last requirement, every element of G has an inverse, and so is one-to-one and onto, that is, a permutation.)

In the modern, axiomatic approach, a group is a set with a binary operation (a function from G^2 to G) which satisfies the closure, associativity, identity and inverse laws. (In fact, closure is not necessary, since a binary operation satisfies it by definition. It remains as a historical relic.)

Of course, the two approaches are essentially the same; we take the modern viewpoint because we do not want to restrict group elements to being transformations of something. Any permutation group is an abstract group with the operation of function composition (since this operation is necessarily associative). Conversely, Cayley's Theorem asserts that any abstract group is isomorphic to a transformation group. Note that, though the axioms (apart from the associative law) match up, the interpretation is different: the identity in an abstract group is defined by its properties with respect to the group operation, and has to be proved unique; the identity in a permutation group is a specific function.

There are two good practical reasons for preferring the modern approach. First, the same group may act as a permutation group on more than one set. This is familiar in parts of combinatorics, such as design theory, where a group of automorphisms of a design acts on both the points and the blocks of the design. Second, we can take the powerful theory of abstract groups (including, for example, the classification of the finite simple groups) and apply its conclusions to permutation groups.

On the other hand, when we are studying a particular permutation group, the work we are doing would not be unfamiliar to mathematicians of the 'classical' period such as Lagrange, Galois, or Jordan.

1.2 Actions and G-spaces

Let Ω be a set. (Since Wielandt's pioneering book [186], it has been customary to use Ω for the set on which permutation groups act, and lower-case Greek letters for its elements.) The *symmetric group* on Ω, written $\mathrm{Sym}(\Omega)$, is the set of all permutations of Ω: it forms a group, with the operation of composition. We write permutations on the right, and compose from left to right: that is, the image of α under the permutation g is αg, and the composition of g and h is gh (so that $\alpha(gh) = (\alpha g)h$). If Ω is a finite set with n elements, we write S_n for the symmetric group on Ω.

A *permutation group* on Ω is a subgroup of $\mathrm{Sym}(\Omega)$.

Let G be a group. An *action* of G on Ω is a homomorphism ϕ from G to $\mathrm{Sym}(\Omega)$. We usually abbreviate $\alpha(g\phi)$ (the image of α under the permutation corresponding to ϕ) to αg.

The image of an action of G on Ω is a permutation group, called the permutation group *induced* on Ω by G, and written G^Ω. We say that an action is *faithful* if its kernel is the identity. If this holds, then G is isomorphic to G^Ω.

There is another way to formalise this concept, in axiomatic terms. A *G-space* is a set Ω with a function $\mu : \Omega \times G \to \Omega$ satisfying the conditions

(A1) $\mu(\mu(\alpha, g), h) = \mu(\alpha, gh)$ for all $\alpha \in \Omega$ and $g, h \in G$;

(A2) $\mu(\alpha, 1) = \alpha$ for any $\alpha \in \Omega$, where 1 is the identity of G.

This is essentially the same thing. For, if ϕ is an action of G on Ω, then the function μ defined by $\mu(\alpha, g) = \alpha(g\phi)$ makes Ω into a G-space. Conversely, suppose that (Ω, μ) is a G-space. For each $g \in G$, there is a function π_g : $\Omega \to \Omega$ given by $\alpha\pi_g = \mu(\alpha, g)$. Then π_1 is the identity function, and $\pi_{g^{-1}}$ is the inverse function to π_g, so each function π_g is a permutation; and then (A1) shows that the map $\phi : g \mapsto \pi_g$ is a homomorphism, that is, an action of G.

Accordingly, when there is no risk of confusion, we will write $\mu(\alpha, g)$ simply as αg.

Each point of view leads to slightly different language for describing the situation. Thus, when we define the property of transitivity in the next section, we will say that G is a transitive permutation group, or that G acts transitively on Ω, or that Ω is a transitive G-space, all meaning the same thing.

We say that two G-spaces Γ and Δ are *isomorphic* if there is a bijection $\theta : \Gamma \to \Delta$ such that, for any $\alpha \in \Gamma$ and $g \in G$, we have

$$\mu_\Gamma(\alpha, g)\theta = \mu_\Delta(\alpha\theta, g).$$

In other words, the diagram of Figure 1.1 commutes, where g on the left refers

$$\begin{array}{ccc} \Gamma & \xrightarrow{\theta} & \Delta \\ g_\Gamma \downarrow & & \downarrow g_\Delta \\ \Gamma & \xrightarrow{\theta} & \Delta \end{array}$$

Figure 1.1: G-space isomorphism

to the action on Γ, and on the right to the action on Δ. Informally, the two sets Γ and Δ can be matched up in such a way that the actions of G on them are 'the same'.

As is usual in algebra, we regard isomorphic G-spaces as being the same. In the next two sections, we will give a classification of G-spaces up to isomorphism.

Finally, we define the *degree* of a permutation group, group action, or G-space to be the cardinality of the set Ω (a finite or infinite cardinal number).

1.3 Orbits and transitivity

Let Ω be a G-space. We define a relation $\sim$ on Ω by the rule that $\alpha \sim \beta$ if and only if there is an element $g \in G$ with $\alpha g = \beta$. Now $\sim$ is an equivalence relation. (The reflexive, symmetric and transitive laws come directly from the identity, inverse and closure laws in the 'old' definition of a permutation group.) The equivalence classes of $\sim$ are the *orbits* of G; and we say that G is *transitive* (or Ω is a *transitive* G-space) if there is just one orbit. Note that each orbit is a G-space in its own right. So we have our first structure theorem:

Theorem 1.1 *Every G-space can be uniquely expressed as a disjoint union of transitive G-spaces.*

From the point of view of permutation groups, however, the situation is not so simple. Suppose that G is a permutation group on Ω, with orbits Ω_i

for $i \in I$, where I is some index set for the set of orbits. Then G acts on each set Ω_i, and so induces a transitive permutation group G^{Ω_i}. These groups G^{Ω_i}, for $i \in I$, are called the *transitive constituents* of G. How is G built from its transitive constituents?

Let $(G_i : i \in I)$ be a family of groups. We define the *cartesian product* $G = \prod_{i \in I} G_i$ to be the set of all functions f from I to $\bigcup_{i \in I} G_i$ which have the property that $f(i) \in G_i$ for all $i \in I$; the group operation is defined 'coordinatewise', that is, for $f_1, f_2 \in G$, we define $f_1 f_2$ by the rule that $(f_1 f_2)(i) = f_1(i) f_2(i)$ for each $i \in I$.

There is a projection map $\phi_i : G \to G_i$ for each $i \in I$, which maps each function f to the element $f(i) \in G_i$. (This construction generalises the *direct product* of finitely many groups.) We say that a subgroup H of G is a *subcartesian product* of the groups $(G_i : i \in I)$ if the restriction of ϕ_i maps H onto G_i for each $i \in I$. Then we have:

Theorem 1.2 *Any permutation group is a subcartesian product of its transitive constituents.*

Example. Consider the two permutation groups G_1 and G_2 on the set $\{1, 2, 3, 4\}$ given by

$$G_1 = \{(1), (1\ 2)(3\ 4)\},$$
$$G_2 = \{(1), (1\ 2), (3\ 4), (1\ 2)(3\ 4)\}.$$

Each has two orbits, namely $\{1, 2\}$ and $\{3, 4\}$; in each case, both the transitive constituents are cyclic groups of order 2. We see that G_2 is the full cartesian product of these two cyclic groups, whereas G_1 is a proper subcartesian product.

1.4 Transitive groups and coset spaces

Let H be a subgroup of G. The (right) *coset space* $H \backslash G$ is the set of right cosets $\{Hx : x \in G\}$, with action of G given by $\mu(Hx, g) = Hxg$ for all $x, g \in G$. (The action is written formally in the μ form here since, with our less formal convention, it would read $Hxg = Hxg$!) It is a transitive G-space. Note that $H \backslash G$ is not the same thing as the set-theoretic difference $H \setminus G$, though typographically they are almost identical. The notation suggests right cosets, with elements of G placed on the right of H. The set of left cosets would be written G/H. Some authors write $G : H$ or $(G : H)$ indiscriminately for either coset space.

The classification of transitive G-spaces, up to isomorphism, is given by the following theorem. If G acts on Ω and $\alpha \in \Omega$, the *stabiliser* of α is the set

$$\{g \in G : \alpha g = \alpha\}$$

of all group elements for which the corresponding permutation fixes α. It is a subgroup of G. If Ω is the coset space $H\backslash G$, then the stabiliser of the *coset* $H1 = H$ is the *subgroup* H.

Theorem 1.3 *(a) Let Ω be a transitive G-space. Then Ω is isomorphic to the coset space $H\backslash G$, where $H = G_\alpha$ for $\alpha \in \Omega$.*

(b) Two coset spaces $H\backslash G$ and $K\backslash G$ are isomorphic if and only if H and K are conjugate subgroups of G.

Sketch proof. (a) The isomorphism is given by

$$\beta \in \Omega \leftrightarrow \{g \in G : \alpha g = \beta\},$$

the set on the right being a right coset of G_α.

(b) The conjugates of G_α are the stabilisers of the points of Ω: this proves the reverse implication. For the forward implication, let θ be an isomorphism between the coset spaces $H\backslash G$ and $K\backslash G$. If θ maps H to Kx, then the stabiliser of H (which is just H) is equal to the stabiliser of Kx (which is $x^{-1}Kx$, a conjugate of K).

So the transitive G-spaces are classified by the conjugacy classes of subgroups of G.

We note two important consequences.

Corollary 1.4 *Let G act transitively on Ω, and $\alpha \in \Omega$.*

(a) A subset S of G is a set of right coset representatives for G_α in G if and only if S contains just one element mapping α to β for all $\beta \in \Omega$.

(b) If G is finite, then

$$|G| = |\Omega| \cdot |G_\alpha|.$$

The second part of the corollary is a form of Lagrange's Theorem, since any subgroup H of G is a point stabiliser in some transitive action of G (namely, on the coset space $G\backslash H$).

From the perspective of permutation groups, once again, the situation is a bit more complicated. The *core* of the subgroup H of G is given by

$$\mathrm{Core}_G(H) = \bigcap_{x \in G} x^{-1}Hx;$$

it is the largest normal subgroup of G which is contained in H. Now the conjugates $x^{-1}Hx$ are the stabilisers of the points Hx of the coset space $H\backslash G$; so $\mathrm{Core}_G(H)$ is the kernel of the action of G on the coset space. In other words, given a group G, the transitive permutation groups isomorphic to G (that is, the faithful transitive actions) are classified by the conjugacy classes of *core-free* subgroups (those whose core is the identity).

Example. Let G be the Klein group $C_2 \times C_2$. How many isomorphism types of n-element G-spaces are there?

The group G has one subgroup of order 4, three of order 2, and one of order 1. Each is conjugate only to itself, since G is abelian. So there are just five transitive G-spaces up to isomorphism, with cardinalities $1, 2, 2, 2, 4$.

Let a_n be the total number of n-element G-spaces, up to isomorphism. Then a_n is the number of ways of paying a bill of n cents with a supply of 1-cent coins, three different kinds of 2-cent coins, and 4-cent coins. By standard combinatorial arguments, we have

$$\sum_{n\geq 0} a_n x^n = \frac{1}{(1-x)(1-x^2)^3(1-x^4)}.$$

Now the right-hand side is analytic in $\mathbb{C}$ apart from poles at 1, -1, i, and $-i$. So the asymptotic form of a_n can be found by standard analytic arguments.

Clearly the same process can be performed (in principle) for any finite group G.

1.5 Sylow's Theorem

We will illustrate the ideas of group actions by proving Sylow's Theorem, the most important result in elementary finite group theory. This was originally proved using arguments with double cosets; the proof given here, based on group actions, is due to Wielandt [182].

Theorem 1.5 (Sylow's Theorem) *Let G be a group of order $n = p^a m$, where p is prime and doesn't divide m. Then*

(a) G has a subgroup of order p^a;

(b) all subgroups of order p^a are conjugate;

(c) any subgroup of p-power order is contained in a subgroup of order p^a.

Proof. To show (a), we let Ω be the set of all subsets of G of cardinality p^a. Then Ω is a set of cardinality $\binom{n}{p^a}$. We define an action of G on Ω by right multiplication: $\mu(S, g) = Sg$ for $S \in \Omega$, $g \in G$. Then Ω is the union of the G-orbits of this action. Note that each G-orbit, as a set of subsets of G, covers all of G: for if $x \in S$, then $y \in Sx^{-1}y$. So there are two types of orbit:

Type 1: orbits containing exactly $n/p^a = m$ sets, covering G without overlapping;

Type 2: orbits containing more than m sets.

Now if S is in a Type 1 orbit, then its stabiliser has order $n/m = p^a$, and so is a subgroup P of the required order. (In this case, if S is chosen to contain the identity, then $S = P$, and the orbit of S is just the set of right cosets of P.) On the other hand, if S is in a Type 2 orbit Δ, then $|\Delta| > m$, so $|\Delta|$ is divisible by p. So, if we can show that $|\Omega|$ is not divisible by p, then it will follow that there must be Type 1 orbits, and (a) will be proved.

It follows from Lucas' Theorem on congruence of binomial coefficients (or can be proved directly) that $\binom{p^a m}{p^a} \equiv m \pmod{p}$. So the number of sets in Type 1 orbits is congruent to m mod p, and the number of subgroups of order p^a is congruent to 1 mod p.

Alternatively, this can be proved with a trick. Consider the cyclic group of order n. This has exactly one subgroup of order p^a, so for this group the number of sets in Type 1 orbits is congruent to m mod p^a. It follows that $\binom{p^a m}{p^a} \equiv m \pmod{p}$, as required.

We prove (b) and (c) together by using a different action of G, by conjugation on the set Ξ of subgroups of order p^a. We need the following observation. Let $|P| = p^a$ and $|Q| = p^b$ $(b \leq a)$. If Q fixes P in this action, then $Q \leq P$. For Q normalises P, and so PQ is a subgroup, of order $|P| \cdot |Q|/|P \cap Q|$. This is a power of p, and hence (by Lagrange's Theorem) cannot be larger than $|P| = p^a$. So $|P \cap Q| \geq |Q|$, whence we have $Q \leq P$. In particular, a subgroup of order p^a can fix only one element of Ξ, namely itself.

We see that P has one orbit of size 1 on Ξ (itself), and all other orbits have size divisible by p. Thus, $|\Xi| \equiv 1 \pmod{p}$, in agreement with our observation in the proof of (a). But we also see that one G-orbit (the one containing P) has size congruent to 1 mod p, and all the others (if any) have size divisible by p. But this argument applies for any subgroup of order p^a, and shows that any such subgroup lies in the unique orbit of size congruent to 1 mod p. So there is only one orbit: in other words, all subgroups of order p^a are conjugate. We have proved (b).

Now Q certainly has a fixed point P in Ξ, since $|\Xi| \equiv 1 \pmod{p}$. By the above remarks, $Q \leq P$. Thus (c) is proved.

1.6 Regular groups

The permutation group G is *semiregular* if only the identity has a fixed point; it is *regular* if it is transitive and semiregular. By the structure theorem for transitive groups (Theorem 1.3), if G is regular on Ω, then Ω is isomorphic to the space of right cosets of the identity; each such coset is a 1-element set, so we can identify the coset space with the set G, on which G acts by right multiplication: $\mu(x, g) = xg$. This is the *right regular representation* of G, which occurs in the proof of Cayley's Theorem.

There is also a *left regular representation* of G on itself, given by the rule that $\mu(x,g) = g^{-1}x$. (The inverse is required here to make a proper action: note that $h^{-1}(g^{-1}x) = (gh)^{-1}x$.) It is, as the name says, also a regular action, and so must be isomorphic to the right regular action: indeed the map $x \mapsto x^{-1}$ is an isomorphism.)

The left and right regular representations commute with each other:

$$g^{-1}(xh) = (g^{-1}x)h.$$

(Note how the associative law for G translates into the commutative law for these actions.) In fact, it can be shown that the centraliser, in the symmetric group, of the (image of the) right regular representation is the left regular representation.

So we have an action of $G^* = G \times G$ on G, the product of the two regular actions:

$$\mu(x,(g,h)) = g^{-1}xh.$$

This transitive action contains both the left and the right regular actions as normal subgroups. The stabiliser of the identity is the *diagonal subgroup*

$$\{(g,g) : g \in G\},$$

acting by conjugation on G. For this reason we call G^* a *diagonal group*.

Example. Higman, Neumann and Neumann [92] constructed an infinite group G with the property that all the non-identity elements of G are conjugate. (Such a group, of course, is simple.) Then the diagonal group G^* has the property that the stabiliser of the identity is transitive on the non-identity elements. As we will see, this means that G is *2-transitive*. Yet it is the direct product of two regular subgroups.

We outline the proof, which depends on the so-called *HNN-construction*. Let G be a group, A and B subgroups of G, and $\phi : A \to B$ an isomorphism. Then the group

$$G' = \langle G, t : t^{-1}at = a\phi \text{ for all } a \in A\rangle$$

has the properties

- G is embedded isomorphically in G';
- any element of G' with finite order lies in a conjugate of G (so that, if G is torsion-free, so is G');
- $t^{-1}Gt \cap G = B$ and $tGt^{-1} \cap G = A$.

In particular, if G is torsion-free and a, b are non-identity elements of G, then G can be embedded in a torsion-free group G' in which a and b are conjugate.

Repeating this process for each pair of non-identity elements of G, we see that, if G is countable and torsion-free, then G can be embedded in a group $G^\dagger$ such that any two non-identity elements of G are conjugate in $G^\dagger$; and $G^\dagger$ is also countable and torsion-free.

Now put $G_0 = G$ and $G_{n+1} = G_n^\dagger$ for all $n \in \mathbb{N}$, and $G^* = \bigcup_{n\in\mathbb{N}} G_n$. Then G^* is countable; and any two non-identity elements of G^* lie in G_n for some n, so are conjugate in G_{n+1} (and *a fortiori* in G^*).

1.7 Groups with regular normal subgroups

Let G be a permutation group on Ω with a regular normal subgroup N. Then there is a bijection from Ω to N, which takes the given action of N on Ω to its action on itself by right multiplication: pick $\alpha \in \Omega$ and set

$$n \in N \leftrightarrow \beta = \alpha n \in \Omega.$$

It turns out that this map is also an isomorphism between the given action of G_α on Ω and its action on the normal subgroup N by conjugation: for, if $g \in G_\alpha$ maps β to γ, and $\alpha n_1 = \beta$, $\alpha n_2 = \gamma$, then

$$\alpha(g^{-1}n_1g) = \alpha n_1 g = \beta g = \gamma,$$

and since $g^{-1}n_1g \in N$, and N is regular, we have $g^{-1}n_1g = n_2$, as required.

Since $G = NG_1$ and $N \cap G_1 = 1$, we have that G is a *semidirect product* of N by G_1.

Conversely, if N is any group, and H a subgroup of its automorphism group $\mathrm{Aut}(N)$, then the semidirect product of N by H acts as a permutation group on N, with N as regular normal subgroup and H as the stabiliser of the identity: the element hn of $N \rtimes H$ acts on N by the rule

$$\mu(x, hn) = x^h n$$

for $x, n \in N$, $h \in H$.

Example 1. If $Z(N) = 1$, then $N \cong \mathrm{Inn}(N)$, the group of inner automorphisms of G, and the semidirect product $N \rtimes \mathrm{Inn}(N)$ is the diagonal group $N^* = N \times N$ described earlier. (Note that $Z(N) = 1$ is a necessary and sufficient condition for the action of N^* to be faithful: see Exercise 1.6.)

Example 2. The group $N \rtimes \mathrm{Aut}(N)$ is the *holomorph* of N.

Example 3. Let V be a vector space, N its additive group, and $H = \mathrm{GL}(V)$, the *general linear group* (the group of all invertible linear transformations of V). Then $N \rtimes H$ is the *affine general linear group* $\mathrm{AGL}(V)$.

1.8 Multiple transitivity

Let k be a positive integer less than $|\Omega|$. We say that G is *k-transitive* on Ω if it acts transitively on the set of all k-tuples of distinct elements of Ω (where the action is *componentwise*: $(\alpha_1, \ldots, \alpha_k)g = (\alpha_1 g, \ldots, \alpha_k g)$).

For $k > 1$, G is k-transitive on Ω if and only if

- G is transitive on Ω, and
- G_α is $(k-1)$-transitive on $\Omega \setminus \{\alpha\}$.

(A special case of this result was used in the example in Section 1.6.)

Note that a k-transitive group is l-transitive for all $l \leq k$. The symmetric group S_n is n-transitive, while the *alternating group* A_n (the set of elements of S_n with even parity) is $(n-2)$-transitive. This is because there are exactly two permutations in S_n which map a given $(n-2)$-tuple of distinct points to another; and they differ by a transposition of the remaining two points, so one is even and one is odd. (See Section 1.17 for a discussion of parity.)

We now investigate when groups with regular normal subgroups can be k-transitive.

Theorem 1.6 *Let G be k-transitive and have a regular normal subgroup N. Assume that $G \neq S_k$.*

(a) If Ω is finite and $k = 2$, then N is an elementary abelian p-group for some prime p.

(b) If $k \geq 3$, then N is an elementary abelian 2-group.

Proof. If G is k-transitive, then G_α acts as a group of automorphisms of N, which is $(k-1)$-transitive on the non-identity elements.

(a) Suppose that G is finite and $k = 2$. Take $x_1, x_2 \in N \setminus \{1\}$. Then some element of G_1 conjugates x_1 to x_2, so these elements have the same order. This order must be prime, since if x has composite order rs then x^r has order s. So N is a p-group. Now $Z(N) \neq 1$, and since no automorphism can map an element of $Z(N)$ to one outside, we must have $Z(N) = N$, and N is elementary abelian.

This fails for infinite groups, as the Higman–Neumann–Neumann diagonal example shows.

(b) Now suppose that $k = 3$, and $|N| > 3$ (possibly infinite). We claim that every element $x \in N$ satisfies $x^2 = 1$. Suppose that $x^2 \neq 1$, and choose $y \neq 1, x, x^2$. Then no automorphism can map x to x and x^2 to y, contrary to assumption. So N is an elementary abelian 2-group.

Note that the affine group $\mathrm{AGL}(V)$ is always 2-transitive, and that it is 3-transitive if V is a vector space over the field with 2 elements.

1.9 Primitivity

Let G be transitive on Ω. A *congruence* is a G-invariant equivalence relation on Ω. A *block* is a subset Δ of Ω such that $\Delta g = \Delta$ or $\Delta g \cap \Delta = \emptyset$ for all $g \in G$. Any congruence class is a block (trivially), and any non-empty block is a congruence class (since its images under G are also blocks and so are pairwise disjoint).

There are two trivial congruences which occur for any group: the relation of equality; and the 'universal' relation with $\alpha \sim \beta$ for all α, β. (Regarded as a set of ordered pairs, the latter is just $\Omega \times \Omega$.) They are the finest and coarsest equivalence relations on Ω. We say that G is *imprimitive* if it has a non-trivial congruence; *primitive* otherwise. (Note that, as defined here, only a transitive group may have the property of being primitive!)

The next result summarises some properties of primitivity.

Theorem 1.7 *(a) A 2-transitive group is primitive.*

(b) A non-trivial normal subgroup of a primitive group is transitive.

(c) The transitive group G on Ω is primitive if and only if G_α is a maximal subgroup of G.

Proof. (a) The only binary relations which are preserved by a 2-transitive group are the empty relation, equality, inequality and the universal relation; only the second and fourth are equivalence relations.

(b) Let N be a non-trivial normal subgroup of the transitive group G. Since $(\alpha N)g = (\alpha g)N$ for any α and g, the orbits of N are blocks.

(c) Suppose that $G_\alpha < H < G$. Then the H-orbit of α is a block. Conversely, if Δ is a non-trivial block (congruence class) containing α, then the setwise stabiliser G_Δ lies strictly between G_α and G.

1.10 Wreath products

We construct the wreath product of two permutation groups. Let H and K be permutation groups on the sets Γ and Δ respectively. Let $\Omega = \Gamma \times \Delta$. Think of Ω as a fibre bundle over the set Δ, with projection map $(\gamma, \delta) \mapsto \delta$; each fibre $\Gamma_\delta = \{(\gamma, \delta) : \gamma \in \Gamma\}$ for fixed $\delta \in \Delta$ is isomorphic to Γ. (See Figure 1.2.)

Let B be the cartesian product of $|\Delta|$ copies of H, one acting on each fibre as H acts on Γ. Thus B is the set of functions from Δ to H with pointwise operations; the action is given by

$$\mu((\gamma, \delta), f) = (\gamma f(\delta), \delta).$$

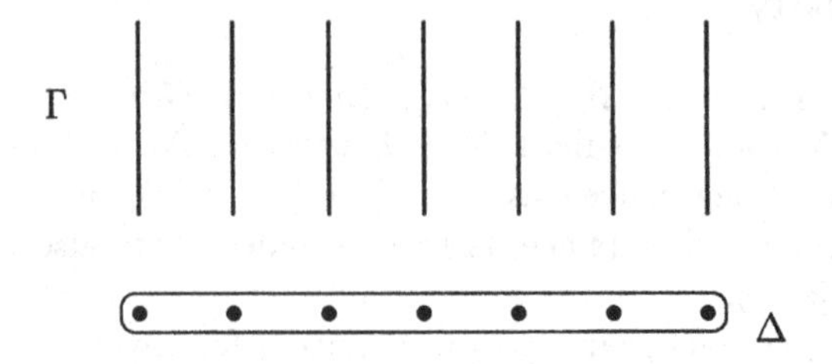

Figure 1.2: A fibre bundle

Let T be a copy of the group K permuting the fibres:

$$\mu((\gamma,\delta),k) = (\gamma,\delta k).$$

Then T normalises B, so the product BT (which is the semidirect product) is a group. This group is the *wreath product* of the permutation groups H and K, written $H \operatorname{Wr} K$.

If H is transitive on Γ and K is transitive on Δ, then $H \operatorname{Wr} K$ is transitive on Ω. If $|\Gamma|, |\Delta| > 1$, it is imprimitive: the fibres are blocks, or (said otherwise) the relation $\sim$, defined by $(\gamma,\delta) \sim (\gamma',\delta')$ if and only if $\delta = \delta'$, is a non-trivial congruence.

The wreath product plays a role for imprimitive groups similar to that of the cartesian product for intransitive groups:

Theorem 1.8 *Let G be transitive but imprimitive on Ω. Let Γ be a non-trivial congruence class, and $H = (G_\Gamma)^\Gamma$ the group induced on Γ by its setwise stabiliser. Let Δ be the set $\{\Gamma g : g \in G\}$ of translates of Γ, and $K = G^\Delta$ the group induced on Δ by G. Then there is a bijection between Ω and $\Gamma \times \Delta$ which embeds G into the wreath product $H \operatorname{Wr} K$.*

The point of this theorem is that, just as we used the orbit decomposition to reduce the study of arbitrary groups to that of transitive groups (modulo difficulties about subcartesian products), so we can further reduce imprimitive groups to 'smaller' ones (modulo difficulties caused by the fact that G isn't the full wreath product). Indeed, if Ω is finite, we can continue this process until all the transitive building blocks are primitive. They are called the *primitive components* of G.

Unfortunately, even in the finite case, the primitive components are not uniquely determined; even their number depends on the way we make the decomposition. There is no Jordan–Hölder Theorem for primitive components!

For example, let $G = \mathrm{AGL}(V)$, where V is a 2-dimensional vector space over a field with q elements, and let G act on the set of *flags* (incident point–line pairs) of the affine plane based on V. (A *line* is a coset of a 1-dimensional subspace of V; two lines are *parallel* if they are cosets of the same subspace.) The lattice of congruences is as shown: numbers on the edges indicate the number of classes of the finer congruence in each class of the larger.

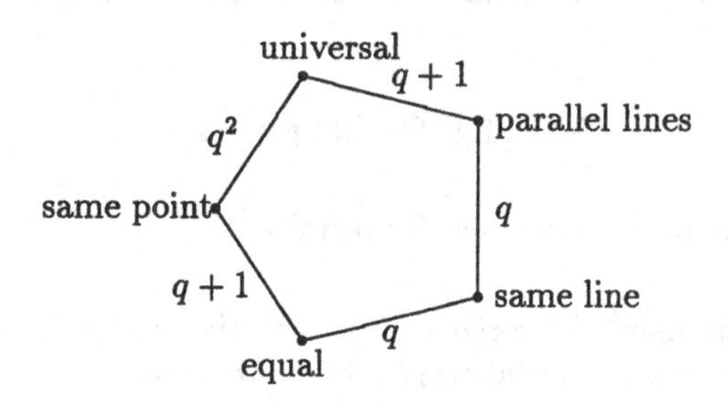

Figure 1.3: Congruences

1.11 Orbitals

Let G be a permutation group on Ω. An *orbital* of G is an orbit of G on the set $\Omega \times \Omega$. The number of orbitals is the *rank* of G.

The *orbital graph* associated with an orbital Δ is the directed graph with vertex set Ω and edge set Δ: in other words, there is a directed edge from α to β, for each $(\alpha, \beta) \in \Delta$.

The orbital *paired* with Δ is

$$\Delta^* = \{(\beta, \alpha) : (\alpha, \beta) \in \Delta\}.$$

We say that Δ is *self-paired* if $\Delta^* = \Delta$. If Δ is self-paired, then its orbital graph can be regarded as an undirected graph.

If G is transitive, there is one *trivial* or *diagonal* orbital, given by $\{(\alpha, \alpha) : \alpha \in \Omega\}$. The corresponding orbital graph has one loop at each vertex. None of the other orbital graphs have loops.

Also, if G is transitive, there is a natural bijection between the orbitals of G and the orbits of a point stabiliser G_α: to the orbital Δ corresponds the G_α-orbit

$$\{\beta \in \Omega : (\alpha, \beta) \in \Delta\}$$

(the set of points β such that there is an edge from α to β in the corresponding orbital graph). The G_α-orbits are called *suborbits*, and their cardinalities are the *subdegrees* of G.

Example. The group $G = C_4$, acting regularly, has rank 4; the four orbital graphs are shown in Figure 1.4.

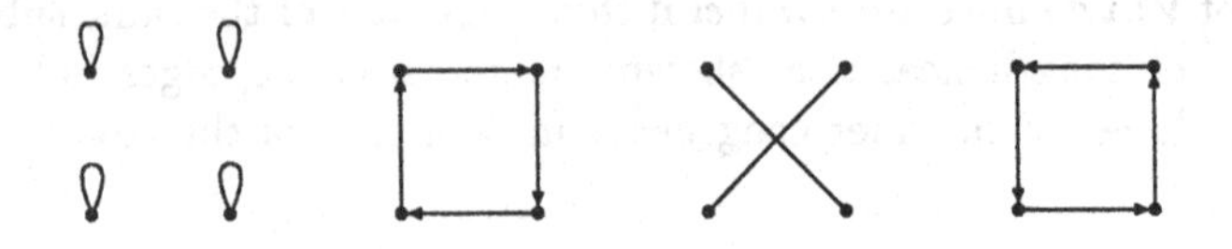

Figure 1.4: Orbital graphs

The orbital graphs give us a test for primitivity:

Theorem 1.9 (Higman's Theorem) *The transitive group G is primitive if and only if all non-trivial orbital graphs for G are connected.*

Proof. A connected component for any orbital graph is a block; conversely if there is a non-trivial congruence $\sim$ with $\alpha \sim \beta$, $\alpha \neq \beta$, then every connected component of the orbital graph for the orbital containing (α, β) is contained in a congruence class of $\sim$.

Remark. There are two concepts of connectivity for directed graphs. The graph is *connected* if it is possible to get from any vertex to any other by a path in the graph, ignoring directions; if this is possible obeying directions, the graph is *strongly connected.* This is the difference between walking and driving in a town with one-way streets. However, we have:

Theorem 1.10 *Any finite vertex-transitive graph (in particular, any orbital graph for a finite transitive permutation group) which is connected is strongly connected.*

Proof. Let $B(\alpha)$ be the set of points which can be reached by directed paths from α. Clearly, if $\beta \in B(\alpha)$, then $B(\beta) \subseteq B(\alpha)$. But there is an element $g \in G$ with $\alpha g = \beta$; and $B(\alpha)g = B(\beta)$. So $B(\alpha)$ and $B(\beta)$ have the same cardinality, and are necessarily equal, whence $\alpha \in B(\beta)$. So, if it is possible to walk from one vertex to another, then it is possible to drive.

This is false in the infinite case: consider the infinite directed path (an orbital graph for the infinite cyclic group acting regularly, see Figure 1.5). Wielandt [184] introduced the concept of *strong primitivity* for infinite transitive groups; it turns out that G is strongly primitive if and only if every orbital graph is strongly connected.

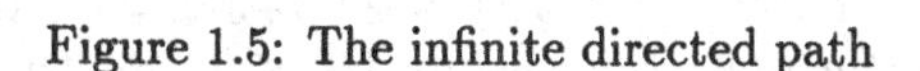

Figure 1.5: The infinite directed path

1.12 Sharp k-transitivity

Let k be a positive integer with $k \leq |\Omega|$. We say that G is *sharply k-transitive* on Ω if, given any two k-tuples $(\alpha_1, \ldots, \alpha_k)$ and $(\beta_1, \ldots, \beta_k)$ of distinct elements of Ω, there is exactly one element $g \in G$ mapping the first k-tuple to the second. In other words, G acts regularly on the set of k-tuples of distinct elements.

Sharp 1-transitivity is just regularity, and any group has such an action. There are many sharply k-transitive groups for $k = 2$ and $k = 3$; they often turn out to be related to geometric objects such as affine and inversive planes. The symmetric group S_n is both sharply n-transitive and sharply $(n-1)$-transitive, and the alternating group A_n is sharply $(n-2)$-transitive (as we showed in fact when we proved it was $(n-2)$-transitive). However, further possibilities are limited by the following theorem of Tits [175], in which no finiteness is assumed:

Theorem 1.11 *The only sharply 4-transitive groups are S_4, S_5, A_6, and the Mathieu group M_{11}.*

Proof. We show that the degree n is finite, and that $n \leq 11$, and $n \leq 6$ if n is even. The cases $n = 7$ and $n = 9$ are ruled out in Exercise 1.18. It is a little more work to prove that there is a unique group for $n = 11$.

Let $\Omega = \{1, 2, 3, 4, \ldots\}$. Let t be the unique element of G which satisfies $t = (1\ 2)(3)(4)\cdots$ (that is, $1t = 2$, $2t = 1$, $3t = 3$, $4t = 4$). Note that $t^2 = 1$, since t^2 fixes at least the four points $1, 2, 3, 4$. Now let H be the subgroup

$$H = \{g \in G : 1g = 1, tg = gt\}$$

of G. Now t fixes two or three points (at least the two points 3 and 4, and at most three since $t \neq 1$), and H permutes among themselves these fixed points; moreover, it acts faithfully on them, since an element of H fixing them all would have at least four fixed points (including 1 and 2). So $|H| \leq 6$.

Now let C be the set of 2-cycles of t other than $(1\ 2)$. We claim that H permutes C transitively, and the stabiliser of an element of C in H has order at least 2.

Let

$$t = (1\ 2)(3)(4)\cdots(i\ j)\cdots(k\ l)\cdots.$$

Take $g \in G$ with $1g = 1$, $2g = 2$, $ig = k$, $jg = l$ (there is a unique such element). Then $g^{-1}tg = (1\ 2) \cdots (k\ l) \cdots$; so $g^{-1}tg = t$, and $g \in H$ maps $(i\ j)$ to $(k\ l)$ as required. Taking $k = j$ and $l = i$, we see that at least one non-identity element of H fixes the cycle $(i\ j)$.

We conclude that $|C| \leq 3$, and so $|\Omega| \leq 2+3+3\cdot 2 = 11$. In particular, Ω is finite. If $|\Omega|$ is even, then t fixes two points, $|H| \leq 2$, and $|\Omega| \leq 2+2+2\cdot 1 = 6$.

Remark. From this, the classification of sharply k-transitive groups for $k > 4$ can be completed. It can be shown that M_{11} can be transitively extended once, to the sharply 5-transitive Mathieu group M_{12}, but no further; so the only such groups are S_k, S_{k+1}, A_{k+2}, and (for $k = 5$) M_{12}.

Remark. The theorem was generalised by Yoshizawa [190], who proved that there is no infinite 4-transitive group in which the stabiliser of any four points is finite. Another generalisation will be given in Theorem 4.22.

Remark. The finite sharply k-transitive groups for $k = 2, 3$ were determined in the 1930s by Zassenhaus [192, 191]. We give a brief survey.

A *near-field* is an algebraic structure with two operations, addition and multiplication, satisfying all of the axioms for a field with the possible exception of the commutative law for multiplication ($xy = yx$) and the left distributive law ($x(y+z) = xy + xz$). Given a near-field F, the *affine transformations* $x \mapsto xa + b$, for $a, b \in F$, $a \neq 0$, form a sharply 2-transitive group. Any finite sharply 2-transitive permutation group G on Ω arises in this way. For it can be shown, by elementary arguments not requiring Frobenius' Theorem, that G has an abelian regular normal subgroup A (see Exercise 1.16). Then we build a near-field with additive group A and multiplicative group G_α, by simply identifying Ω with the regular group A (so that α corresponds to 0), and $\Omega \setminus \{\alpha\}$ with the regular group G_α, in the usual way.

Thus, determining the finite sharply 2-transitive groups is equivalent to determining the finite near-fields. This was achieved by Zassenhaus [192]. Most of them are *Dickson near-fields*, which arise from fields by a procedure involving 'twisting' the multiplication by field automorphisms; their multiplicative groups are metacyclic. (Exercise 1.19 outlines the construction.) In addition, there are just seven *exceptional near-fields*, with orders 5^2, 7^2, 11^2 (twice), 23^2, 29^2, and 59^2. In the exceptional near-fields, the multiplicative group has a normal subgroup isomorphic to $\mathrm{SL}(2,3)$ or $\mathrm{SL}(2,5)$. See Dixon and Mortimer [64] for details of all this.

The finite sharply 3-transitive groups are more restricted. For any prime power q, the group $\mathrm{PGL}(2,q)$ is sharply 3-transitive. Apart from these, there is one further example for each q which is an odd square. In this case, the outer automorphism group of $\mathrm{PSL}(2,q)$ contains a Klein group $\{1, a, b, c\}$, where $\mathrm{PSL}(2,q)\{1,a\} = \mathrm{PGL}(2,q)$ and b is induced by the field automorphism of order 2. Then $\mathrm{PSL}(2,q)\{1,c\}$ is also sharply 3-transitive. For $q = 9$, it is this group which is the point stabiliser in M_{11} (sometimes called M_{10}); the group $\mathrm{P\Sigma L}(2,9) = \mathrm{PSL}(2,9)\{1,b\}$ is isomorphic to S_6, acting on the ten partitions of $\{1,\ldots,6\}$ into two 3-sets, and is not 3-transitive.

The depth of our ignorance about infinite sharply 2-transitive groups is shown by the fact that it is not even known whether or not every such group has a regular normal subgroup.

See Clay [51] for a survey of the general topic of near-rings, and Passman [146] or Dixon and Mortimer [64] for the classification of finite sharply 2- and 3-transitive groups.

1.13 The Schreier–Sims algorithm

Often we are given a permutation group G on a finite set $\Omega = \{1,2,\ldots,n\}$ by means of a set of generating permutations $g_1,\ldots,g_n$. The *Schreier–Sims algorithm* is a technique for getting information about G from such data.

Step 1. We find the orbit of α by repeatedly applying generators until no new points are obtained. This is very similar to the algorithm for finding connected components of a graph.

For example, suppose that $n = 4$, $g_1 = (1\ 2)(3\ 4)$, and $g_2 = (1\ 2\ 3)$. Starting from the point 1, we find that $2 = 1g_1$, $3 = 2g_2$, $4 = 3g_1$, and then we have everything.

We record this in a vector whose β entry is the generator that was applied to reach β for the first time. In the α position, we just put some marker to indicate that it is the 'base point' of the orbit. In the above example, the vector would be $(*,1,2,1)$.

Note that from this data we can easily recover a set of coset representatives for G_α in G: in the above case, the coset representatives mapping 1 to $1,2,3,4$ are 1, g_1, g_1g_2, $g_1g_2g_1$ respectively.

If there are several orbits, we can use distinct negative numbers for the base points of distinct orbits. In this way, a single vector of integers (the *Schreier vector*) contains complete information about the orbits and coset representatives of stabilisers.

Step 2. We now find generators for G_α. These are obtained from *Schreier's Lemma*:

Theorem 1.12 *Let $G = \langle g_1,\ldots,g_r\rangle$. Let H be a subgroup of index n in G, with coset representatives $x_1,\ldots,x_n$, where $x_1 = 1$ is the representative of H. Let $\overline{g}$ be the representative of the coset Hg. Then*

$$H = \langle x_i g_j (\overline{x_i g_j})^{-1} : 1 \le i \le n, 1 \le j \le r\rangle.$$

The proof will be given at the end of this section.

Recursion. Now apply these two steps recursively, until the group we reach is trivial (that is, all its generators are the identity). At this point, we end up with

(a) a *base*, a sequence $(\alpha_1, \alpha_2, \ldots, \alpha_m)$ whose pointwise stabiliser is the identity; and

(b) a *strong generating set*, a set $S = S_1 \cup \ldots \cup S_m$ of group elements such that S_i is a set of coset representatives for $G(i)$ in $G(i-1)$ for $1 \le i \le m$, where $G(i)$ is the pointwise stabiliser of $(\alpha_1, \ldots, \alpha_i)$.

From this we can easily find

(c) the order of G: for each i, we have $|S_i| = |G(i-1) : G(i)|$, and $G(0) = G$, $G(m) = 1$, so

$$|G| = |G(0) : G(1)| \cdots |G(m-1) : G(m)| = |S_1| \cdots |S_m|;$$

and

(d) a membership test for G. Given any permutation x of Ω, first check whether $\alpha_1 x$ lies in the G-orbit of α_1, that is, whether it is equal to $\alpha_1 s_1$ for some $s_1 \in S_1$. If it is not, then $x \notin G$. If it is, then $x \in G$ if and only if $x s_1^{-1} \in G(1)$, which can be checked recursively. We end up with the situation that *either*

- we demonstrate that $x \notin G$; *or*
- we find $s_i \in S_i$ for $i = 1, \ldots, m$ so that $x s_1^{-1} \cdots s_m^{-1} \in G(m) = 1$, whence

$$x = s_m s_{m-1} \cdots s_1 \in G.$$

Note that, in (d), if $x \in G$, then x is expressed as an element of $S_m S_{m-1} \cdots S_1$, and this expression is unique. In particular,

(e) the set S really does generate G;

(f) we have another proof of the formula of (c) for the order of G;

(g) we can choose a random element of G (with all elements equally likely), by choosing random elements of $S_m, S_{m-1}, \ldots, S_1$ and multiplying them.

Proof of Schreier's Lemma. First we note that all the proposed generators belong to H, since each has the form $g\overline{g}^{-1}$ with $Hg = H\overline{g}$.

Next, we calculate the inverses of these elements. Consider the element $x_i g_j (\overline{x_i g_j})^{-1} = x_i g_j x_k^{-1}$, where $H x_i g_j = H x_k$. Its inverse is then $x_k g_j^{-1} x_i^{-1}$, where $Hx_kg_j^{-1} = Hx_i$. So the inverses are expressions of similar form but replacing generators of G by their inverses.

Now we show that these elements generate H. Take an arbitrary element $h \in H$. We can write

$$h = g_{j_1}^{\epsilon_1} g_{j_2}^{\epsilon_2} \cdots g_{j_m}^{\epsilon_m},$$

with $\epsilon_k = \pm 1$. Rewrite this as

$$h = x_{i_0} g_{j_1}^{\epsilon_1} x_{i_1}^{-1} \cdot x_{i_1} g_{j_2}^{\epsilon_2} x_{i_2}^{-1} \cdots x_{i_{m-1}} g_{j_m}^{\epsilon_m} x_{i_m}^{-1} \cdot x_{i_m},$$

where $x_{i_0} = 1$ (the coset representative of H) and, for each k, $x_{i_k} = \overline{x_{i_{k-1}} g_{j_k}^{\epsilon_k}}$. Now an easy induction shows that $x_{i_m} = \overline{h}$; because $h \in H$, we have $x_{i_m} = 1$, and we have expressed h as a product of the supposed generators.

The proof is complete.

1.14 Jerrum's filter

There is a serious practical problem with the Schreier–Sims algorithm as presented in the last section. If we are given r generators for G, then we obtain roughly rn generators for $G(1)$, rn^2 generators for $G(2)$, and so on. Ultimately, when we reach the identity, all the generators will be the identity, and they can be discarded. But we may run into serious memory problems storing all the intermediate generators if, say, n is in the thousands.

The way to deal with this is to process the generators as they are produced by Schreier's Lemma to obtain a smaller set. One way to do this is *Jerrum's filter* [101], which enables us to keep the number of generators bounded.

Theorem 1.13 *Any subgroup of S_n can be generated by at most $n-1$ elements. Moreover, such a generating set can be found 'on-line', in the sense that if S is a suitable generating set for H and g any element of S_n, then it is possible to find a suitable generating set S' for $\langle H, g\rangle$.*

Proof. Let $\Omega = \{1, 2, \ldots, n\}$. With any non-identity permutation g of Ω, we associate an element and a 2-subset of Ω as follows:

- $\imath(g)$ is the smallest point of Ω moved by g;
- $e(g) = \{\imath(g), \imath(g)g\}$.

Now, given any set S of permutations, the set $e(S) = \{e(g) : g \in S\}$ is the edge set of a graph on the vertex set Ω. We claim that

> *any subgroup of S_n can be generated by a subset S such that the graph $e(S)$ contains no cycles.*

This will prove the theorem, since an acyclic graph on n vertices has at most $n-1$ edges (with equality if and only if it is a tree). So suppose that $e(S)$ is acyclic and g is any permutation. We want an 'acyclic' generating set S' for $\langle S, g\rangle$. There are three cases:

- $g = 1$. Then take $S' = S$.
- $e(S \cup \{g\})$ is acyclic. Take $S' = S \cup \{g\}$.
- $e(S \cup \{g\})$ contains a unique cycle, which includes $e(g)$. Let $S_1 = S \cup \{g\}$. Moreover, for any set T of permutations containing a unique cycle, we let $m(T) = \sum_{g \in T} i(g)$. We show how to construct from S_1 a set S_2 with $\langle S_1\rangle = \langle S_2\rangle$ such that either $e(S_2)$ is acyclic, or $e(S_2)$ contains a unique cycle and $m(S_2) > m(S_1)$. Since $m(T)$ is bounded above for any such set (for example, by n^2, since there are at most n permutations and $i(g) \leq n$ for any $g \neq 1$), the second alternative can only occur finitely many times, and eventually we reach a set S' such that $\langle S'\rangle = \langle S_1\rangle$ and $e(S')$ is acyclic. Take this as the required S'.

It remains to show how to do the replacement step above.

Let C be the unique cycle in $e(S_1)$, and let i be the smallest point lying on any edge of C. Then we can travel round the cycle starting at i, recording our progress by elements g or g^{-1} according as the edge is from $i(g)$ to $i(g)g$ or *vice versa*. We obtain a product $h = g_{i_1}^{\epsilon_1} \cdots g_{i_k}^{\epsilon_k}$, with $\epsilon_j = \pm 1$ for $1 \leq j \leq k$, such that h fixes i. Clearly it also fixes every point smaller than i. So, if we delete from S_1 the element g_{i_1} and replace it with h, we increase the value of m (since $i(h) > i(g_{i_1}) = i$). Moreover, removing g_{i_1} produces an acyclic set, and so the addition of h at worst creates one cycle; and g_{i_1} can be expressed in terms of h and the other generators, so the groups generated by the two sets are equal.

This concludes the proof.

Remark. The result is not best possible. McIver and Neumann [124] showed that, for $n > 3$, any subgroup of S_n can be generated by at most $\lfloor n/2 \rfloor$ elements. This is best possible: consider the elementary abelian 2-group generated by $\lfloor n/2 \rfloor$ disjoint transpositions. However, no algorithmic proof (and in particular, no on-line proof) of this bound seems to be known.

1.15 The length of S_n

Another attempt to bound the number of generators of a permutation group for purposes of computational group theory, due to Babai [8], used the notion of the length of S_n.

The *length* $l(G)$ of a group G is the length l of the longest chain

$$1 = G_0 < G_1 < \cdots < G_l = G$$

of subgroups of G. Now, if H is a subgroup of G, then $l(H) \leq l(G)$ (simply take a longest subgroup chain for H, and add the term G at the end if $H \neq G$). Also, a group G can be generated by at most $l(G)$ elements (on taking g_i to be any element of $G_i \setminus G_{i-1}$ for $i = 1, \ldots, l(G)$). So the length $l(S_n)$ is an upper bound for the number of generators of any subgroup of S_n.

The length of S_n has been calculated: see the next theorem, due to Cameron, Solomon and Turrull [47]. The bound for number of generators, however, is less good than that resulting from Jerrum's filter in the last section.

Theorem 1.14

$$l(S_n) = \left\lceil \frac{3n}{2} \right\rceil - b(n) - 1,$$

where $b(n)$ is the number of ones in the base 2 expansion of n.

Sketch proof. We have to construct a chain of this length, and show that no longer chain is possible.

The occurrence of $b(n)$ in the formula suggests how to construct a chain. If n is not a power of 2, write it as a sum of distinct powers of 2, say $n = 2^{a_1} + \cdots + 2^{a_k}$, where $k = b(n)$. If $k > 1$, we can descend in $k - 1$ steps to the group $S_{2^{a_1}} \times \cdots \times S_{2^{a_k}}$, and we have to handle each of the factors. If $n = 2^a$ with $a > 1$, then take the two steps $S_{2m} > S_m \operatorname{Wr} S_2 > S_m \times S_m$, where $m = 2^{a-1}$, and then proceed with the factors. It is an exercise to show that the length of the chain is as claimed (Exercise 1.23).

Note that other, completely different, chains of the same length are possible for some small symmetric groups.

The proof that no longer chain is possible depends (rather weakly) on the Classification of Finite Simple Groups, and will be discussed in Chapter 4.

Remark. If N is a normal subgroup of G, then $l(G) = l(N) + l(G/N)$. So the length of a group is the sum of the lengths of its composition factors. So it suffices to compute $l(G)$ for simple groups G. The value of $l(A_n)$ follows from the theorem above. Other simple groups have also been investigated.

1.16 At the keyboard

The Schreier–Sims algorithm is only one of many sophisticated algorithms that have been devised for computing with permutation groups, enabling us to answer questions out of the range of hand calculation. These algorithms have been gathered together into programming systems in which a group (or other algebraic structure) can be manipulated as freely as a number or matrix in a more traditional programming language. The most popular systems are MAGMA and GAP.

In the remainder of this section, I will describe a session with GAP [161]. The object of this exercise is to construct the unique example of a 2-transitive permutation group whose unique minimal normal subgroup is simple but not 2-transitive.

The group in question is $\mathrm{P\Sigma L}(2,8)$, obtained by adjoining a field automorphism to $\mathrm{PSL}(2,8)$. As an example of the use of the system for computing the group generated by some explicit permutations, I will construct generators for this group with bare hands, rather than finding the group in one of the extensive GAP libraries of groups.

On the back of an envelope, we construct the field GF(8); its elements are of the form $a\alpha^2 + b\alpha + c$, where $a, b, c \in \mathrm{GF}(2) = \{0, 1\}$ and α is a primitive 7th root of unity, satisfying $\alpha^3 = \alpha + 1$. Thus the successive powers of α are

$$\begin{aligned}
\alpha^0 &= 1, \\
\alpha^1 &= \alpha, \\
\alpha^2 &= \alpha^2, \\
\alpha^3 &= \alpha + 1, \\
\alpha^4 &= \alpha^2 + \alpha, \\
\alpha^5 &= \alpha^2 + \alpha + 1, \\
\alpha^6 &= \alpha^2 + 1.
\end{aligned}$$

We number these elements as $1, 2, 3, 4, 5, 6, 7$ respectively, and use 8 to denote the zero element of the field.

The group $\mathrm{PSL}(2,8)$ acts on the projective line over GF(8), which can be regarded as $\mathrm{GF}(8) \cup \{\infty\}$. We use 9 to denote the point ∞ on the projective line. So $\mathrm{PSL}(2,8)$ will be a subgroup of the symmetric group on $\{1, \ldots, 9\}$.

Having started up GAP, we define the symmetric group as follows:

```
gap> S:=SymmetricGroup(9);
Group( (1,9), (2,9), (3,9), (4,9), (5,9), (6,9), (7,9), (8,9) )
```

Now $\mathrm{PSL}(2,8)$ can be generated by the three permutations $x \mapsto x + 1$, $x \mapsto \alpha x$, and $x \mapsto 1/x$. These act on $\{1, \ldots, 9\}$ as follows:

```
gap> g1:=(1,8)(2,4)(3,7)(5,6);
(1,8)(2,4)(3,7)(5,6)
gap> g2:=(1,2,3,4,5,6,7);
(1,2,3,4,5,6,7)
```

```
gap> g3:=(2,7)(3,6)(4,5)(8,9);
(2,7)(3,6)(4,5)(8,9)
```

Now we let N be the group they generate, and check its order, which should be $9.8.7 = 504$.

```
gap> N:=Subgroup(S,[g1,g2,g3]);
Subgroup( Group( (1,9), (2,9), (3,9), (4,9), (5,9), (6,9),
                 (7,9), (8,9) ),
          [ (1,8)(2,4)(3,7)(5,6), (1,2,3,4,5,6,7),
            (2,7)(3,6)(4,5)(8,9) ] )
gap> Size(N);
504
```

We are interested in the group $G = \mathrm{P\Sigma L}(2,8)$, which can now be obtained as the normaliser of $\mathrm{PSL}(2,8)$ in S_9:

```
gap> G:=Normalizer(S,N);
Subgroup( Group( (1,9), (2,9), (3,9), (4,9), (5,9), (6,9),
                 (7,9), (8,9) ),
          [ (1,8)(2,4)(3,7)(5,6), (1,2,3,4,5,6,7),
            (2,7)(3,6)(4,5)(8,9), (3,5,7,6,8,9,4),
            (4,6,9)(5,8,7) ] )
gap> Size(G);
1512
```

We want to compute the action of G on 28 points, which can be taken as the cosets of the normaliser of a Sylow 3-subgroup of G. So we compute the Sylow subgroup, its normaliser, and the action of G on the set of right cosets. This is done as follows. (As you will see if you try this out for yourself, when we define a new group, GAP gives us a list of generators; this is not necessary to see what is happening so I will suppress most of the output in future. If you don't want to see this output, end each command with two semicolons, as shown.)

```
gap> P:=SylowSubgroup(G,3);;
gap> Q:=Normalizer(G,P);;
gap> GG:=Operation(G,RightCosets(G,Q),OnRight);;
```

The output of the last command (if viewed) enables us to see that the group we have called `GG` is indeed a permutation group on 28 points.

We want to check that G is 2-transitive. For this, we check that it is transitive, and that the stabiliser of a point is transitive on the other 27 points.

```
gap> Orbits(GG,[1..28]);
[ [ 1, 13, 18, 28, 10, 14, 17, 6, 27, 24, 3,
      26, 22, 23, 9, 15, 4, 11, 21, 2,
      16, 12, 7, 25, 8, 19, 5, 20 ] ]
gap> GGa:=Stabilizer(GG,1);;
gap> Orbits(GGa,[1..28]);
[ [ 1 ], [ 2, 22, 27, 6, 11, 7, 17, 24, 12,
      10, 9, 5, 13, 18, 19, 8, 3, 15,
      25, 20, 16, 26, 23, 14, 21, 4, 28 ] ]
```

By inspection, the orbit lengths of the group and stabiliser are 28 and 27 respectively, so it is indeed 2-transitive.

Now the minimal normal subgroup is most easily obtained as the derived subgroup.

```
gap> NN:=DerivedSubgroup(GG);;
gap> Orbits(NN,[1..28]);
[ [ 1, 25, 18, 2, 24, 7, 10, 21, 5, 22, 15,
      20, 14, 9, 17, 11, 4, 26, 23, 6,
      12, 19, 27, 16, 28, 3, 13, 8 ] ]
gap> NNa:=Stabilizer(NN,1);;
gap> Orbits(NNa,[1..28]);
[ [ 1 ], [ 2, 15, 11, 16, 25, 24, 7, 21, 18 ],
  [ 3, 13, 28, 8, 14, 6, 23, 9, 27 ],
  [ 4, 17, 26, 20, 22, 5, 12, 10, 19 ] ]
```

We see that, while PSL(2, 8) is transitive, the stabiliser has three orbits of length 9, so it is not 2-transitive. However, we check that it is primitive:

```
gap> IsPrimitive(NN,[1..28]);
true
```

And, just as a check:

```
gap> DisplayCompositionSeries(GG);
<G> (3 gens, size 1512)
 | Z(3)
<S> (2 gens, size 504)
 | A(1,8)=L(2,8)~B(1,8)=O(3,8)~C(1,8)=S(2,8)~2A(1,8)=U(2,8)
<1> (0 gens)
```

So GG has a simple normal subgroup of index 3 which GAP recognizes by the several names listed (including L(2,8), which means PSL(2, 8)).

For more details about GAP, see the homepage at

http://www-gap.dcs.st-and.ac.uk/~gap/

Some of the more sophisticated algorithms for computing in permutation groups built into GAP are discussed in the book by Seress [165].

1.17 Appendix: Cycles and parity

The *cycle decomposition* of a permutation g is essentially the orbit decomposition of Ω under the cyclic group $\langle g \rangle$. We write $(\alpha_1\ \alpha_2\ \dots\ \alpha_m)$ for the cycle

$$\alpha_1 \mapsto \alpha_2 \mapsto \cdots \mapsto \alpha_m \mapsto \alpha_1$$

of g. An infinite cycle extends in both directions; we write it as a sequence indexed by the integers, $(\dots\ \alpha_{-1}\ \alpha_0\ \alpha_1\ \dots)$, say, if g maps α_i to α_{i+1} for all i.

We see that, if there are only finitely many cycles of g (and in particular, if Ω is finite), then g is the product of its cycles. In this assertion, we can ignore cycles of length 1. We often omit them from the notation as well, assuming that every point of Ω not mentioned is fixed by g.

In the remainder of this section, we assume that Ω is finite, say $\Omega = \{1, 2, \dots, n\}$. The *cycle structure* of an element of S_n is the list of the lengths of its cycles (arranged in non-increasing order).

Theorem 1.15 *(a) Two elements of S_n are conjugate if and only if they have the same cycle structure.*

(b) An element of S_n which has a_i cycles of length i for $1 \le i \le n$ (where $\sum ia_i = n$) has centraliser of order z, and lies in a conjugacy class of order $n!/z$, where

$$z = 1^{a_1} a_1!\ 2^{a_2} a_2!\ \cdots\ n^{a_n} a_n!\,.$$

Sketch proof. Part (a) follows from the fact that we obtain the cycle decomposition of $x^{-1}gx$ from that of g simply by replacing each point α by its image under x. (For, if $\alpha g = \beta$, then $(\alpha x)(x^{-1}gx) = \beta x$.)

For (b), write out the cycles of g leaving blank spaces for the points of Ω. These points can be inserted in $n!$ ways. But each permutation with the given cycle structure arises z times: for each cycle can start at any of its points, and cycles of the same length can be written in any order.

The number of permutations with a given number of cycles is an 'unsigned' Stirling number. More precisely, the *Stirling numbers of the first kind*, $s(n, k)$, are defined by the rule that $(-1)^{n-k}s(n, k)$ is the number of elements of S_n

which have k cycles (including cycles of length 1). It can be shown that these numbers have a particularly simple generating function:

$$\sum_{k=1}^{n} s(n,k)x^k = x(x-1)(x-2)\cdots(x-n+1).$$

We define the *parity* of an element $g \in S_n$ to be the parity of $n - c(g)$, where $c(g)$ is the number of cycles of g. (The parity of g is regarded as an element of $\mathbb{Z}/(2)$.) The parity can be written as $\sum(l(C)-1)$, summed over the cycles C of g; it is enough to take cycles of length greater than 1 in this expression. Hence the parity of an element $g \in S_n$ does not change if we regard g as a permutation of $\{1,\ldots,N\}$ for $N > n$.

Substituting $x = 1$ in the generating function for the Stirling numbers, we see that, if $n \geq 2$, then

$$\sum_{k=1}^{n} s(n,k) = 0,$$

so there are equally many (namely $n!/2$) elements with even and odd parity. There is a more algebraic reason for this, which we now explore.

How is the parity of a permutation affected by multiplying it by a transposition (a permutation interchanging two points and fixing the rest)? There are two cases. If the transposition t interchanges points in different cycles of g, say

$$g = (\alpha_1 \ \ldots \ \alpha_r)(\beta_1 \ \ldots \ \beta_s)\cdots, \qquad t = (\alpha_1 \ \beta_1),$$

then we find that

$$gt = (\alpha_1 \ \ldots \ \alpha_r \ \beta_1 \ \ldots \ \beta_s)\cdots;$$

the two cycles of g are 'stitched together' in gt. Dually, if the points interchanged by t lie in the same cycle of g, then this cycle is cut into two cycles of gt. In either case, the number of cycles has changed by 1, and so the parity has changed.

Theorem 1.16 *(a) Any element in S_n is a product of transpositions.*

(b) For $n \geq 2$, parity is a homomorphism from S_n onto $\mathbb{Z}/(2)$.

Proof. (a) It is enough to prove this for a single cycle. Check that

$$(\alpha_1 \ \alpha_2 \ \ldots \ \alpha_m) = (\alpha_1 \ \alpha_2)(\alpha_1 \ \alpha_3)\cdots(\alpha_1 \ \alpha_m).$$

(b) Let $p(g)$ denote the parity of g. Given $g, h \in S_n$, write $h = t_1 \cdots t_r$, where $t_1, \ldots, t_r$ are transpositions. The earlier argument and a simple induction show that

$$p(gh) = p(g) + r \pmod 2.$$

In particular, $p(h) = r \pmod 2$; so $p(gh) = p(g) + p(h)$ in $\mathbb{Z}/(2)$, and p is a homomorphism. For $n \geq 2$, any transposition has odd parity, and p is onto.

We note in passing that $p(g)$ is the parity of the number of transpositions whose product is g (in any expression for g as a product of transpositions).

Corollary 1.17 *For $n \geq 2$, the set of elements of S_n having even parity is a normal subgroup of index 2.*

The subgroup in the corollary is the *alternating group*, written $\mathrm{Alt}(\Omega)$, or A_n if $\Omega = \{1, \ldots, n\}$.

Remark. Sometimes, instead of parity, we talk about the *sign* of a permutation g, given by $\mathrm{sign}(g) = (-1)^{p(g)}$. Then sign is a homomorphism from S_n onto the multiplicative group $\{\pm 1\}$ for $n \geq 2$.

1.18 Exercises

1.1. Let H be a subgroup of the transitive permutation group G. Prove that H is transitive if and only if $G_\alpha H = G$.

1.2. Find all permutation groups of degree 10 which have orbits of length 4 and 6 and act on these orbits as the symmetric group S_4 and the dihedral group D_{12} respectively, where S_4 acts in the natural way and D_{12} acts as on the vertices of a regular hexagon.

1.3. Prove that a nilpotent primitive permutation group is cyclic of prime order. (Don't *assume* finiteness.)

1.4. Let G be the group of permutations of $\{1, 2, \ldots, 12\}$ generated by the two permutations

$$a = (1\ 2)(3\ 4)(5\ 7)(6\ 8)(9\ 11)(10\ 12)$$

and

$$b = (1\ 2\ 3)(4\ 5\ 6)(8\ 9\ 10).$$

Prove that G is transitive. Find coset representatives and Schreier generators for a point stabiliser, and hence show that G is 2-transitive.

[*Note:* In fact G is the Mathieu group M_{12}, and is sharply 5-transitive. The generators come from Grothendieck's *dessins d'enfants* [160]. You may like to continue the algorithm to find the order and the degree of transitivity of G.]

1.5. Let G be a subgroup of the symmetric group S_n, and let H be the centraliser of G in S_n. Of the following four implications,

(a) G is transitive if and only if H is semiregular.

(b) G is semiregular if and only if H is transitive.

three are true and one is false; prove the true ones and find a counterexample to the false one.

Deduce that a transitive abelian group is regular.

1.6. Let G be a group. Consider the *diagonal action* of $G \times G$ on G by the rule

$$\mu(x,(g,h)) = g^{-1}xh.$$

Prove that

(a) the action is faithful if and only if the centre of G is trivial;

(b) the action is primitive if and only if G is simple;

(c) the action is 2-transitive if and only if all non-identity elements of G are conjugate.

1.7. The *affine* and *projective linear groups* are defined as follows.

- $\mathrm{GL}(n,F)$ is the group of invertible linear transformations of the vector space $V = F^n$.
- $\mathrm{AGL}(n,F)$ is the group of permutations of V of the form

$$x \mapsto xA + c : A \in \mathrm{GL}(n,F), c \in V.$$

- $\mathrm{PGL}(n,F)$ is the group of permutations of the 1-dimensional subspaces of V induced by $\mathrm{GL}(n,F)$. It is isomorphic to $\mathrm{GL}(n,F)/Z$, where Z is the group of central scalar transformations (of the form cI, where c is in the centre of F).

We denote $\mathrm{GL}(n,\mathrm{GF}(q))$ by $\mathrm{GL}(n,q)$, and similarly for the other cases.

Prove that

$$\begin{array}{lll} \mathrm{AGL}(1,2) \cong S_2, & \mathrm{AGL}(1,3) \cong S_3, & \mathrm{AGL}(2,2) \cong S_4 \\ \mathrm{PGL}(2,2) \cong S_3, & \mathrm{PGL}(2,3) \cong S_4, & \mathrm{PGL}(2,4) \cong A_5. \end{array}$$

1.8. Let F be a skew field in which all elements different from 0 and 1 are conjugate in the multiplicative group. Prove that

(a) the group

$$G_1 = \{x \mapsto a^{-1}xb + c : a, b, c \in F, a, b \neq 0\}$$

of permutations of F is 3-transitive, and has two sharply 2-transitive normal subgroups;

(b) the group $\mathrm{PGL}(2,F)$ is 4-transitive.

[*Note:* Skew fields with this property were constructed by P. M. Cohn [53].]

1.9. Show that the group of isometries of $\mathbb{R}^n$ is primitive.

1.10. Find the rank, and describe the orbitals, of the symmetric group S_n acting on the set of k-element subsets of $\{1, \ldots, n\}$. Prove that the action is primitive if and only if $n \neq 2k$.

1.11. Suppose that G acts transitively on Γ and Δ. Take points $\gamma \in \Gamma$ and $\delta \in \Delta$. Prove that the numbers of G-orbits on $\Gamma \times \Delta$, of G_γ-orbits on Δ, and of G_δ-orbits on Γ, are all equal.

1.12. Let G be a primitive group of odd order. By the Feit–Thompson theorem [74], G is soluble, and so a minimal normal subgroup N of G is abelian. Show that N acts regularly.

Identify Ω with N as in Section 1.7. Show that the permutation of N given by $n \mapsto n^{-1}$ normalises G.

1.13. Use the results of the preceding question to show that, if G is a finite permutation group which is transitive on unordered 2-sets but is not 2-transitive, then G is a subgroup of index 2 in a 2-transitive group.

1.14. Show that a finite primitive group which contains a transposition is the symmetric group.

1.15. Show that the minimum number of transpositions in S_n whose product is an n-cycle is $n-1$. Show further that the product of $n-1$ transpositions is an n-cycle if and only if the associated edges form a tree. Deduce that a given n-cycle can be written as the product of $n-1$ transpositions in exactly n^{n-2} ways. [*Hint*: On the set $\{1, 2, \ldots, n\}$, there are $(n-1)!$ n-cycles and n^{n-2} trees.]

1.16. Let G be a finite sharply 2-transitive permutation group.

(a) Show that G contains $n(n-2)$ elements fixing one point, and $n-1$ elements with no fixed point.

(b) Let x be a fixed-point-free element in G. Show that the centraliser of x consists of the identity and fixed-point-free elements. Deduce that x has at least $n-1$ conjugates.

(c) Hence show that the fixed-point-free elements and the identity form a regular normal subgroup.

1.17. Show that the holomorph of a cyclic group of prime order p is sharply 2-transitive.

1.18. Let G be sharply 4-transitive of degree $p+2$, with p an odd prime. Show that the normaliser of a Sylow p-subgroup in G has order $2p(p-1)$. Deduce that the centraliser of P contains a transposition, and so G is the symmetric group.

Deduce that there is no sharply 4-transitive group of degree 7 or 9.

1.19. This exercise outlines the construction of Dickson near-fields.

(a) Let F be a skew field, and ϕ a map from the multiplicative group of F to its automorphism group, satisfying the condition

$$(a^{b\phi} \cdot b)\phi = (a\phi) \cdot (b\phi)$$

for all $a, b \neq 0$. Define a new multiplication on F by the rule

$$a \circ b = a^{b\phi} \cdot b$$

for $a, b \neq 0$, with $a \circ 0 = 0 \circ a = 0$ for all a. Prove that $(F, +, \circ)$ is a near-field.

(b) Let q be a prime power, and n a positive integer. Suppose that every prime divisor of n divides $q - 1$, and that, if $q \equiv 1 \pmod 4$, then n is not divisible by 4. Let $m(k) = (q^k - 1)/(q - 1)$. Prove that

- n divides $m(n)$;
- the numbers $m(0) = 0, m(1) = 1, m(2), \ldots, m(n-1)$ form a complete set of residues modulo n.

(c) Let q and n be as in (b). Let θ be the automorphism $x \mapsto x^q$ of $F = \mathrm{GF}(q^n)$. Let ω be a generator of the (cyclic) multiplicative group of F, and let H be the subgroup of index n in the multiplicative group generated by ω^n. Define ϕ by the rule

$$a\phi = \theta^k \text{ if and only if } a \in H\omega^{m(k)},$$

where $k \in \{0, 1, \ldots, n-1\}$. Prove that ϕ satisfies the condition of (a), and so defines a near-field. Prove also that the multiplicative group of this near-field is an extension of a cyclic group of order $(q^n - 1)/n$ by a cyclic group of order n.

1.20. Prove that the stabiliser of a point in the sharply 3-transitive group of degree $q + 1$ different from $\mathrm{PGL}(2, q)$ (constructed in Section 1.12) is the affine group over a Dickson near-field with $n = 2$.

1.21. Prove that no 4-transitive group can have a sharply 3-transitive normal subgroup.

1.22. If N is a normal subgroup of G, show that $l(G) = l(N) + l(G/N)$. Deduce that the length of a group is the sum of the lengths of its composition factors.

1.23. Complete the proof that the chain of subgroups constructed in the proof of Theorem 1.14 has length $\lceil 3n/2 \rceil - b(n) - 1$.

1.24. (a) Show that S_{15} has a subgroup chain of maximal length beginning

$$S_{15} > S_5 \operatorname{Wr} S_3 > S_5 \operatorname{Wr} C_3 > S_5 \times S_5 \times S_5.$$

(b) Show that S_7 has a chain of maximal length beginning

$$S_7 > A_7 > \mathrm{PSL}(3,2).$$

1.25. Let $m(G)$ be the maximum cardinality of a minimal generating set for G (one with the property that any proper subset generates a proper subgroup).

(a) Prove that $m(G) \leq l(G)$, where $l(G)$ is the length of G.

(b) Prove that $m(S_n) \geq n-1$.

(I am grateful to Jan Saxl for this exercise. Julius Whiston (personal communication) has shown that $m(S_n) = n-1$.)

1.26. Prove that any element of A_n is a product of 3-cycles.

The following exercises indicate an inductive approach to the study of multiply transitive groups. Although a definitive result was not obtained, some notable partial results were obtained by Nagao, O'Nan, and others. In these exercises, G denotes a finite permutation group.

1.27. (a) Let G be a transitive permutation group, and H the stabiliser of a point. Prove that $N_G(H)$ acts transitively (and regularly) on the fixed points of H.

(b) More generally, if G is t-transitive and H is the stabiliser of t points, then $N_G(H)$ is sharply t-transitive on the fixed points of H.

(c) Deduce that, if G is 4-transitive, then the stabiliser of four points fixes 4, 5, 6 or 11 points.

Remark. Nagao [136] showed that, if the stabiliser of four points fixes more than four points, then G is sharply 4-transitive.

1.28. (a) Let G be transitive, and P a Sylow subgroup of the stabiliser of a point. Prove that $N_G(P)$ acts transitively on the fixed points of P.

(b) More generally, if G is t-transitive and P is a Sylow subgroup of the stabiliser of t points, then $N_G(P)$ is t-transitive on the set of fixed points of P.

1.29. (a) Let G be a permutation group on the finite set Ω, and let p be a prime. Suppose that, for any $\alpha \in \Omega$, there is a p-subgroup of G whose only fixed point is α. Prove that G is transitive. [*Hint:* Compare the second part of the proof of Sylow's Theorem.]

(b) Let G be transitive on Ω, and p be a prime. Let P be a p-subgroup of G maximal with respect to fixing more than one point. Prove that $N_G(P)$ is transitive on the set of fixed points of P. [*Hint:* If P is a Sylow p-subgroup of G_α, then the result follows from the preceding exercise, so suppose not. Then for any fixed point β of P, a Sylow p-subgroup Q of G_β properly contains P, and so $N_Q(P) > P$. By maximality of P, $N_Q(P)$ fixes only β.]

(c) More generally, let G be t-transitive on Ω, and let p be a prime. Let P be a p-subgroup of G maximal with respect to fixing more than t points of Ω. Prove that $N_G(P)$ is t-transitive on the set of fixed points of P.

Remark. This argument seems to have been discovered independently by Gleason and by Livingstone and Wagner.

1.30. Let H and K be subgroups of G, with $K \leq H$. We say that K is *weakly closed* in H with respect to G if, for all $g \in G$ such that $g^{-1}Kg \leq H$, it holds that $g^{-1}Kg = K$. We say that K is *strongly closed* in H if, for all $g \in G$, we have $g^{-1}Kg \cap H \leq K$.

(a) Prove that, if G is t-transitive and K is weakly closed in the t-point stabiliser with respect to G, then $N_G(K)$ is t-transitive on the fixed point set of K.

(b) Prove that, if G is 2-transitive and K is strongly closed in $G_{\alpha\beta}$ with respect to G_α, then $N_G(K)$ is 2-transitive on the fixed point set of G.

Remark. Part (b) is due to O'Nan [143]. This is more difficult, and you may want to use O'Nan's paper as a crib. Note that, unlike (a), the hypothesis only refers to the proper subgroup G_α of G. It follows that the translates under G of the fixed-point set of K are the lines of a *linear space* (that is, any two points of Ω are contained in a unique such translate). Using this, O'Nan showed that, if G is 2-transitive and G_α has a normal abelian subgroup A which is not semiregular on $\Omega\setminus\{\alpha\}$, then $\mathrm{PSL}(d,q) \leq G \leq \mathrm{P\Gamma L}(d,q)$ for some $d \geq 2$ and some q. For the hypotheses of part (b) are satisfied by $K = A_\beta$, and the linear space can be identified with a projective space.

1.31. This exercise shows an application of the concept of orbital graphs. It strengthens an old theorem of W. A. Manning [133, 134]. This was my own introduction to combinatorial methods in permutation groups [29].

Theorem 1.18 *If G is primitive on Ω, and G_α acts 2-transitively on its largest orbit, then G is 3-transitive.*

Let G be a finite permutation group which is primitive but not doubly transitive. Suppose that G has an orbital Γ with subdegree $|\Gamma(\alpha)| = k$ such that the subconstituent $G_\alpha^{\Gamma(\alpha)}$ is 2-transitive. Show that G has an orbital Δ with subdegree l satisfying $l = k(k-1)/p$ for some integer $p \leq k/2$ (so that, in particular, $l \geq 2(k-1)$). [*Hint:* Consider the case where Γ is self-paired, so that the orbital graph is undirected. Show that G acts transitively on paths of length 2 in the orbital graph, and take Δ to consist of the set of pairs of vertices at distance 2. Then p is the number of common neighbours of two vertices at distance 2. If $p > k/2$, show that the relation of being equal or at distance 2 is an equivalence relation. How should the argument be modified if Γ is not self-paired?]

Remark. Using elementary combinatorial methods, it is possible to show that $p = O(k^{4/5})$ (see [30, II]). However, using the classification of finite simple groups, it can be shown that $p \leq 6$. The extreme case is the Higman–Sims simple group [91], acting on 100 points: the stabiliser of a point is the Mathieu group M_{22}, acting 3-transitively on an orbit of 22 points and with rank 3 on an orbit of 77 points.

1.32. A theorem of Weiss [180] asserts:

Theorem 1.19 *If G is primitive but not regular, then no G_α-orbit except $\{\alpha\}$ has size coprime to that of the largest G_α-orbit.*

Prove this. [*Hint:* If H acts transitively on Γ and Δ, with $(|\Gamma|, |\Delta|) = 1$, then H_γ is transitive on Δ for $\gamma \in \Gamma$.]

Remark. A primitive group can have coprime subdegrees: an example is the sporadic Janko group J_1, which acts primitively on 266 points with rank 5 and subdegrees $1, 11, 12, 110, 132$ (see Livingstone [120]). No primitive group of smaller rank has this property (but the proof requires the classification of finite simple groups).

1.33. Show that, if G is a finite transitive permutation group of odd degree, in which all the subdegrees are odd, then G has odd order. [*Hint:* Show by counting edges that no orbital graph can be undirected.]

1.34. Suppose that $n \geq 5$.

(a) Let $g \in S_n$, $g \neq 1$, and suppose that g is not a fixed-point-free involution. Prove that there is a conjugate h of g in S_n such that gh is a 3-cycle.

(b) Prove that the alternating group A_n can be generated by $\lfloor n/2 \rfloor$ 3-cycles.

(c) Hence show that, if $g \in S_n$, $g \neq 1$, then there exist n conjugates of g which generate S_n (if g is an odd permutation) or A_n (if g is an even permutation). [The case where g is a fixed-point-free involution requires special treatment.]

1.35. Figure 1.6 shows the *seven-point projective plane*, a structure with seven points and seven lines (of which six are shown as straight lines in the figure and one as a circle).

An *automorphism* is a permutation of the points which maps each line to a line.

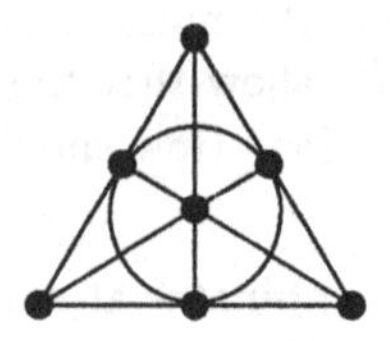

Figure 1.6: The seven-point projective plane

(a) Show that there are 168 automorphisms, forming a 2-transitive group G.

(b) Show that all the automorphisms are even permutations, and that G has index 15 in A_7.

(c) Show that A_7 acts 2-transitively on the cosets of G.

1.36. A subset S of a permutation group G on Ω is *sharply transitive* if and only if, for all $\alpha, \beta \in \Omega$, there is a unique element $g \in S$ such that $\alpha g = \beta$. Show that S is sharply transitive if and only if it is a set of (left or right) coset representatives for G_α for all $\alpha \in \Omega$.

1.37. Define cycle structure for permutations in an infinite symmetric group, and show that two elements of $\mathrm{Sym}(\Omega)$ are conjugate if and only if they have the same cycle structure.

CHAPTER 2

Representation theory

2.1 Historical note

The first edition of Burnside's *Theory of Groups of Finite Order* [28] was published one hundred years ago; it was the first book on group theory in English. In the preface, Burnside explained his decision to treat permutation representations but not matrix representations as follows:

> It may then be asked why, in a book which professes to leave all applications on one side, a considerable space is devoted to substitution groups [permutation groups]; while other particular modes of representation, such as groups of linear transformations, are not even referred to. My answer to this question is that while, in the present state of our knowledge, many results in the pure theory are arrived at most readily by dealing with properties of substitution groups, it would be difficult to find a result that could be most directly obtained by the consideration of groups of linear transformations.

However, in 1911, when the second edition appeared, he wrote,

> Very considerable advances in the theory of groups of finite order have been made since the appearance of the first edition of this book. In particular the theory of groups of linear substitutions has been the subject of numerous and important investigations by several writers; and the reason given in the original preface for omitting any account of it no longer holds good.

The 'several writers' referred to are most notably Frobenius, Schur, and Burnside himself. Among the theorems of pure group theory which had been proved using what we now call representation theory (or character theory) were

- *Frobenius' Theorem*: If G is a finite group with a subgroup H such that $H \cap g^{-1}Hg = 1$ for all $g \notin H$, then G has a normal subgroup N such that $HN = G$, $H \cap N = 1$.
- *Burnside's $p^\alpha q^\beta$ Theorem*: A finite group whose order is divisible by only two primes is soluble.
- *Burnside's Theorem on groups of prime degree*: A transitive permutation group of prime degree is either soluble or 2-transitive.

In fact, Frobenius' Theorem can be expressed in the language of permutation groups, by taking the action of G on $H\backslash G$, as follows:

Theorem 2.1 *If G is a finite transitive permutation group with $G_{\alpha\beta} = 1$ for $\alpha \neq \beta$, then G has a regular normal subgroup.*

See Passman [146] for a proof of Frobenius' Theorem, and for more information, such as a proof that the *Frobenius kernel* (the regular normal subgroup) is nilpotent, and analysis of the structure of the *Frobenius complement* (the point stabiliser).

2.2 Centraliser algebra

Let G be a permutation group on a set Ω, where Ω is finite, say $|\Omega| = n$. The basis of the representation-theoretic approach is that each element $g \in G$ can be represented by a *permutation matrix* $P(g)$, an $n \times n$ matrix with rows and columns indexed by Ω, having (α, β) entry

$$P(g)_{\alpha\beta} = \begin{cases} 1 & \text{if } \alpha g = \beta, \\ 0 & \text{otherwise.} \end{cases}$$

The map $P : g \mapsto P(g)$ is easily checked to be a homomorphism from G to $\mathrm{GL}(n, F)$ (the group of invertible $n \times n$ matrices over F), for any field F.

The *centraliser algebra* of G is the set of all $n \times n$ matrices which commute with all the matrices $P(g)$ for $g \in G$. It is an algebra over F (that is, closed under addition, scalar multiplication and matrix multiplication). A short calculation shows that, for any $g \in G$, the $(\alpha g, \beta g)$ entry of $P(g)^{-1}MP(g)$ is equal to the (α, β) entry of M. So a matrix lies in the centraliser algebra if and only if its entries are constant on the G-orbits on Ω^2 (that is, the *orbitals*).

With each orbital Δ we associate a *basis matrix* $A(\Delta)$, with (α,β) entry

$$A(\Delta)_{\alpha\beta} = \begin{cases} 1 & \text{if } (\alpha,\beta) \in \Delta, \\ 0 & \text{otherwise.} \end{cases}$$

We see that the centraliser algebra is spanned by the basis matrices. But they are linearly independent (since they are zero–one matrices with their ones in disjoint positions). So the dimension of the centraliser algebra of G is equal to the *rank* of G (the number of G-orbits on Ω^2).

The closure under multiplication shows that, if $\Delta_1, \ldots, \Delta_r$ are the orbitals, then we have

$$A(\Delta_i)A(\Delta_j) = \sum_{k=1}^{r} p_{ij}^k A(\Delta_k),$$

for some numbers p_{ij}^k. This formula also has a combinatorial proof which gives an interpretation to these numbers: p_{ij}^k is the number of points γ such that $(\alpha,\gamma) \in \Delta_i$ and $(\gamma,\beta) \in \Delta_j$, where (α,β) is any pair chosen from Δ_k. (This follows from the multiplication rule for zero–one matrices: the (α,β) entry of the product is just the number of points γ for which the (α,γ) entry of the first factor and the (γ,β) entry of the second are both 1.) This shows that the numbers p_{ij}^k are non-negative integers.

The centraliser algebra and the numbers p_{ij}^k will be considered further in the next chapter.

2.3 The Orbit-Counting Lemma

This result is sometimes called 'Burnside's Lemma', but is not due to Burnside: it was certainly known to Frobenius. It is the basis of the Redfield–Pólya–de Bruijn theory of enumeration.

Let $\mathrm{fix}(g)$ be the number of points of Ω which are fixed by the element $g \in G$.

Theorem 2.2 *Let G be a permutation group on the finite set Ω. Then the number of orbits of G on Ω is equal to the average number of fixed points of an element of G, that is,*

$$\frac{1}{|G|}\sum_{g\in G} \mathrm{fix}(g).$$

Proof. Consider the following bipartite graph. The vertex set is $\Omega \cup G$, and there is an edge from $\alpha \in \Omega$ to $g \in G$ just if $\alpha g = \alpha$.

We use a standard combinatorial trick: we count in two different ways the edges of Ω. If a point $\alpha \in \Omega$ lies in a G-orbit Δ, then the number of edges passing through α is $|G_\alpha| = |G|/|\Delta|$; so the number of edges through all the points of Δ is $|G|$, and the total number of edges is $m|G|$, where m is the number of orbits. On the other hand, looking at G, we see that the number of edges is $\sum_{g \in G} \text{fix}(g)$. The theorem is proved.

Remark. It is a slogan of modern enumeration theory that the ability to count a set is closely related to the ability to pick a random element from that set (with all elements equally likely). The above simple proof of the Orbit-Counting Lemma is related to a Markov chain method devised by Jerrum [102] for choosing a random orbit of a permutation group. It works as follows. (See Grimmett and Stirzaker [85] for details on Markov chains.) We start with a point $\alpha \in \Omega$, chosen from any probability distribution whatever. One step of the Markov chain involves the following:

- move to a random neighbour of α, that is a random element g of G fixing α;
- move to a random neighbour of g, that is, a random point β fixed by g.

Since it is possible to move from any point α to any other point β in a single step (via the identity of G), the Markov chain is *irreducible* and *aperiodic*. So it has a unique limiting distribution, which is approached by iterating the procedure from any starting point.

We *claim* that, in the limiting distribution, *all orbits are equally likely*, that is, the probability of reaching α is inversely proportional to the size of the orbit containing α (or, directly proportional to $|G_\alpha|$). An informal proof of this assertion follows: you should try to write out a formal proof.

We start with a 'distribution' on Ω, a function ϕ given by $\phi(\alpha) = |G_\alpha|$. (This is not a probability distribution, since it is not normalised to have sum 1.) In the first part of the Markov step, we distribute $\phi(\alpha)$ equally over the neighbours of α. As in the above proof, each edge carries one unit; so the total received by g is $f(g) = \text{fix}(g)$. Now in the second step, we distribute $f(g)$ equally over the neighbours of g. Again each edge carries one unit, and so the amount received by α is $|G_\alpha| = \phi(\alpha)$. This shows that the function ϕ is proportional to the limiting distribution of the Markov chain.

Of course, the usefulness of the method depends on how rapidly the Markov chain converges to its limiting distribution. A Markov chain is said to be *rapidly mixing* if it approaches its limiting distribution exponentially fast (as measured by *variation distance*). Goldberg and Jerrum [81] have constructed examples where the above Markov chain is not rapidly mixing, though it is so in some important cases. We do not tackle such questions here.

Example. The method is particularly valuable when a relatively small group acts on a very large set. There are many examples of this (see Jerrum's cited paper). Here is one.

A *labelled graph* on n vertices is a graph on the vertex set $\{1,2,\ldots,n\}$. The number of such graphs is $2^{n(n-1)/2}$, and there is a simple procedure for selecting one at random: we choose, independently with probability 1/2, for each pair of vertices, whether or not to join those vertices by an edge. (We regard the vertices as being labelled by the first n natural numbers; two labelled graphs are counted as 'the same' if and only if there is a bijection between them which preserves not only the graph structure but also the labelling of the vertices.)

An *unlabelled graph* on n vertices is an isomorphism class of n-vertex graphs. (This time we regard the vertices as being indistinguishable, and two graphs are 'the same' if and only if there is a bijection between them preserving the graph structure.) How do we choose a random unlabelled graph? We observe that an unlabelled graph is an orbit of the symmetric group S_n on the set of labelled graphs. Since S_n is very much smaller than the number of labelled graphs, we use Jerrum's Markov chain.

We examine what the steps of the Markov chain involve. Remember that Ω is the set of all labelled graphs on $\{1,\ldots,n\}$, and $G = S_n$. We start with any graph Γ.

- The stabiliser of Γ in S_n is just the automorphism group $\mathrm{Aut}(\Gamma)$. Computing the automorphism group of a graph is one of a select group of computational problems which are known to lie in the complexity class NP but are not known to be either in P or NP-complete. Some people have speculated that it lies strictly between these two classes. (See Garey and Johnson [79] for a discussion of complexity classes.) However, there are programs such as `nauty` [125], which compute the automorphism group efficiently in practice.

- We need next a random element g of $G = \mathrm{Aut}(\Gamma)$. We saw in Section 1.13 that this can be done easily once a strong generating set has been computed using the Schreier–Sims algorithm: if S_i is a set of coset representatives for $G(i)$ in $G(i-1)$ for $i = 1,\ldots,m$ (where $G(i)$ is the stabiliser of the first i base points), then multiplying random elements of $S_m, S_{m-1},\ldots,S_1$ gives a random element of G. (Note that `nauty` provides a base and strong generating set for $\mathrm{Aut}(\Gamma)$ as its output.)

- Now we need a random graph fixed by g. For this, calculate the cycles of g on the set of 2-element subsets of $\{1,\ldots,n\}$, and for each cycle, choose (independently with probability 1/2) whether the pairs in that cycle are edges or non-edges.

2.4 Character theory

This section and the few following ones give a very brief introduction to character theory of finite groups. For more details, see Curtis and Reiner [57], or another book on the subject.

Let G be a finite group. A *matrix representation* of G (over $\mathbb{C}$) is a homomorphism $M : G \to \mathrm{GL}(n, \mathbb{C})$: that is, $M(gh) = M(g)M(h)$ for $g, h \in G$. The number n is its *degree.*

Two matrix representations M and N are *equivalent* if they are related by a change of basis, that is, $N(g) = X^{-1}M(g)X$ for all $g \in G$, where X is invertible.

The representation M is *reducible* if it is equivalent to a representation of the form

$$\begin{bmatrix} M_1(g) & O \\ * & M_2(g) \end{bmatrix},$$

that is, if there is a proper G-invariant subspace of $\mathbb{C}^n$. It is *irreducible* otherwise. M is *decomposable* if it is equivalent to a representation of the form

$$\begin{bmatrix} M_1(g) & O \\ O & M_2(g) \end{bmatrix},$$

that is, if there is a decomposition $\mathbb{C}^n = U \oplus W$ into proper G-invariant subspaces U and W. It is *indecomposable* otherwise.

Maschke's Theorem (valid for representations of G over any field whose characteristic is zero or not a divisor of $|G|$) asserts that any G-invariant subspace has a G-invariant complement. So reducibility and decomposability are equivalent, and any representation can be decomposed into irreducible ones (in a unique way, by the Krull–Schmidt theorem).

The *character* of the representation M is the function $\chi : G \to \mathbb{C}$ given by $\chi(g) = \mathrm{Trace}\, M(g)$ for $g \in G$.

Theorem 2.3 *(a) Any character is constant on the conjugacy classes of G.*

(b) Two representations are equivalent if and only if they have the same character.

Statement (a) and the forward implication in (b) follow immediately from the fact that $\mathrm{Trace}\, X^{-1}AX = \mathrm{Trace}\, A$.

By (b), it makes sense to call a character *irreducible* if it is the character of an irreducible representation. Now any character is expressible as a sum of irreducible characters. Also, the *degree* of a character χ is the degree of the corresponding representation; it is equal to $\chi(1)$ (since 1 is represented by the identity matrix).

A representation of degree 1 is the same as its character. If G is abelian, then all its irreducible characters have degree 1. Any group has a character of degree 1, mapping every group element to the complex number 1; this is the *principal character* of the group.

A *class function* is a function from G to $\mathbb{C}$ which is constant on the conjugacy classes. The class functions form a vector space, whose dimension is equal to the number of conjugacy classes. There is a hermitian inner product on this space, given by

$$\langle \phi, \psi \rangle = \frac{1}{|G|} \sum_{g \in G} \phi(g)\overline{\psi(g)}.$$

Theorem 2.4 *The irreducible characters of G form an orthonormal basis for the space of class functions. In particular, the number of irreducible characters is equal to the number of conjugacy classes.*

The *character table* of G is the square matrix K with rows indexed by irreducible characters and columns by conjugacy classes, with (χ_i, C_j) entry equal to $\chi_i(g)$ for $g \in C_j$. The orthonormality of characters gives us that

$$\frac{1}{|G|} \sum_{g \in G} \chi(g)\overline{\phi(g)} = \begin{cases} 1 & \text{if } \chi = \phi, \\ 0 & \text{otherwise,} \end{cases}$$

where χ and ϕ are irreducible characters. This is the *first orthogonality relation.* It can be written as

$$KC\overline{K}^{\top} = |G|I,$$

where C is a diagonal matrix whose elements are the cardinalities of the conjugacy classes. Hence

$$\overline{K}^{\top} K = |G|C^{-1},$$

the right-hand side being a diagonal matrix whose entries are the orders of the centralisers of conjugacy class representatives. (For let G act on itself by conjugation: the orbits are the conjugacy classes, and the stabiliser of g is its centraliser.) Thus,

$$\sum_{\chi \in \mathrm{Irr}(G)} \overline{\chi(g)}\chi(h) = \begin{cases} |C_G(g)| & \text{if } g \text{ and } h \text{ are conjugate,} \\ 0 & \text{otherwise,} \end{cases}$$

the *second orthogonality relation.* (Here $\mathrm{Irr}(G)$ denotes the set of irreducible characters of G.)

2.5 The permutation character

The permutation matrices of Section 2.2 give a matrix representation of G. The trace of the permutation matrix $P(g)$ is just the number of ones on the diagonal of G, which is the number of points fixed by G. So the *permutation character* π is given by

$$\pi(g) = \mathrm{fix}(g)$$

for $g \in G$.

Let

$$\pi = \sum_{\chi \in \mathrm{Irr}(G)} m_\chi \chi$$

be the decomposition of π into irreducible characters. The Orbit-Counting Lemma takes the form

$$\text{number of orbits of } G = \langle \pi, 1 \rangle = m_1,$$

where 1 is the principal character of G. Also, if r is the *rank* of G, the number of orbits on Ω^2, then

$$r = \langle \pi^2, 1 \rangle = \langle \pi, \pi \rangle = \sum_{\chi \in \mathrm{Irr}(G)} m_\chi^2.$$

The first equality holds because the number of ordered pairs fixed by g is $\pi(g)^2$ (an ordered pair is fixed if and only if both components are fixed), and so the permutation character of G on Ω^2 is π^2; the second because π, being real-valued, can be transferred from one side of the hermitian inner product to the other.

In particular, we see that G is 2-transitive if and only if $\pi = 1 + \chi$ for some irreducible character χ.

We saw that the rank is equal to the dimension of the centraliser algebra. This can be seen in a more algebraic way. The key is *Schur's Lemma*:

Theorem 2.5 *Let M be an irreducible matrix representation of G over a field F. Then the centraliser algebra of M is a division algebra over F (a division ring containing F in its centre).*

Proof. Let B be any non-zero matrix in the centraliser algebra. Since $BM(g) = M(g)B$ for all $g \in G$, a short calculation shows that the image and kernel of B are invariant under all $M(g)$. By irreducibility, the image is the whole space F^n, and the kernel is zero; so B is invertible. Thus every element of the centraliser algebra has an inverse, and it is a division algebra.

Since $\mathbb{C}$ is algebraically closed, the only finite-dimensional division algebra over $\mathbb{C}$ is $\mathbb{C}$ itself.

Now take an arbitrary representation M; suppose that it decomposes as the sum of m_i copies of the irreducible representation M_i, for $i = 1, \ldots, k$. That is, M is equivalent to the representation

$$\begin{bmatrix} I_{m_1} \otimes M_1 & O & \ldots & O \\ O & I_{m_2} \otimes M_2 & \ldots & O \\ \vdots & \vdots & \ddots & \vdots \\ O & O & \ldots & I_{m_k} \otimes M_k \end{bmatrix}.$$

The centraliser algebra then consists of all matrices with shape

$$\begin{bmatrix} A_1 \otimes I_{f_1} & O & \ldots & O \\ O & A_2 \otimes I_{f_2} & \ldots & O \\ \vdots & \vdots & \ddots & \vdots \\ O & O & \ldots & A_k \otimes I_{f_k} \end{bmatrix},$$

where A_i is an arbitrary $m_i \times m_i$ matrix, and f_i is the degree of M_i.

Applying this to the permutation character, we see that the rank of G (which is the dimension of the centraliser algebra) is equal to $\sum_{\chi \in \mathrm{Irr}(G)} m_\chi^2$.

Block's Lemma relates the number of orbits in different actions of G. We say that the character ϕ_1 is *contained in* the character ϕ_2 if $\phi_2 = \phi_1 + \psi$ for some character ψ. Also, an *incidence structure* consists of two sets Ω_1 and Ω_2 (whose elements are often called *points* and *blocks* respectively) together with an *incidence relation*, a subset R of $\Omega_1 \times \Omega_2$. The *incidence matrix* of the relation R is the matrix whose rows and columns are indexed by Ω_1 and Ω_2 respectively, with entries 1 on elements of R and 0 elsewhere. It is said to have *full rank* if its rows are linearly independent. For many important classes of incidence structures, such as 2-designs (or balanced incomplete block designs), it is known that the incidence relation has full rank. Finally, an *automorphism* of an incidence structure consists of a permutation of Ω_1 and a permutation of Ω_2 which preserve the incidence relation.

Theorem 2.6 *Let G act on sets Ω_1 and Ω_2. Suppose that the permutation character of G on Ω_1 is contained in the permutation character on Ω_2. Then the number of orbits of G on Ω_2 is at least as great as the number on Ω_1. In particular, this holds if G is a group of automorphisms of an incidence structure whose incidence relation has full rank.*

Proof. Let π_1 and π_2 be the permutation characters. If $\pi_2 = \pi_1 + \psi$ for some character ψ, then

$$\langle \pi_2, 1 \rangle = \langle \pi_1, 1 \rangle + \langle \psi, 1 \rangle \geq \langle \pi_1, 1 \rangle.$$

Suppose that G is a group of automorphisms of an incidence structure whose incidence matrix M has full rank. Then there exist invertible matrices

X and Y such that $X^{-1}MY = [I\ O]$. (This is because we can reduce M to the form $[I\ O]$ by elementary row and column operations.) Suppose that P_1 and P_2 are the permutation matrix representations. Since any $g \in G$ is an automorphism, we have

$$P_1(g)M = MP_2(g).$$

Then

$$X^{-1}P_1(g)X \cdot X^{-1}MY = X^{-1}MY \cdot Y^{-1}P_2(g)Y.$$

Putting

$$Y^{-1}P_2(g)Y = \begin{bmatrix} A(g) & B(g) \\ C(g) & D(g) \end{bmatrix},$$

we find that $A(g) = X^{-1}P_1(g)X$ and $B(g) = O$. Thus P_2 is equivalent to a decomposable representation containing P_1 as a component. Taking traces, we find $\pi_2 = \pi_1 + \psi$, where $\psi(g) = \mathrm{Trace}\, D(g)$ is a character, as required.

2.6 The diagonal group

Recall the diagonal action of $G^* = G \times G$ on G, given by the rule that $\mu(x,(g,h)) = g^{-1}xh$. By considering the permutation character of G^*, we can throw more light on the character theory of G.

First we compute the permutation character of G^*. We have $g^{-1}xh = x$ if and only if $x^{-1}gx = h$. If g and h are not conjugate, there are no such x; if they are conjugate, the set of such x is a right coset of the centraliser of g. (Consider G acting on itself by conjugation.) So

$$\pi((g,h)) = \begin{cases} |C_G(g)| & \text{if } g \text{ and } h \text{ are conjugate,} \\ 0 & \text{otherwise.} \end{cases}$$

Now any irreducible character of $G \times G$ is of the form $\chi(g,h) = \phi(g)\psi(h)$ for some $\phi, \psi \in \mathrm{Irr}(G)$. (If M and N are matrix representations with characters ϕ and ψ respectively, then χ is the trace of the representation $(g,h) \mapsto M(g) \otimes N(h)$, where $\otimes$ denotes the Kronecker product of matrices. The irreducibility of this representation is easily shown by computing that $(\chi,\chi) = 1$; and the number of characters of this form is equal to the number of conjugacy classes in $G \times G$, so we have them all.) We denote this character by $[\phi,\psi]$. By the second orthogonality relation (see page 41), we see that

$$\pi = \sum_{\phi \in \mathrm{Irr}(G)} [\phi, \overline{\phi}]$$

is the decomposition of π into irreducible characters of G^*.

This illustrates that the rank of G^* (which, we know, is equal to the number of conjugacy classes of G) is equal to the number of irreducible characters of G.

Putting $g = h$, we recover the second orthogonality relation

$$|C_G(g)| = \sum_{\phi \in \mathrm{Irr}(G)} \phi(g)\overline{\phi(g)}.$$

In particular, with $g = 1$:

Theorem 2.7

$$|G| = \sum_{\phi \in \mathrm{Irr}(G)} \phi(1)^2.$$

The *regular character* ρ_G of G is the character of the regular representation, given by

$$\rho_G(g) = \begin{cases} |G| & \text{if } g = 1, \\ 0 & \text{otherwise.} \end{cases}$$

Now putting $h = 1$ in G^* gives the regular representation of G. We obtain

$$\rho_G = \sum_{\phi \in \mathrm{Irr}(G)} \phi(1)\phi,$$

that is, the multiplicity of any irreducible character in ρ_G is equal to its degree.

2.7 Frobenius–Schur index

Schur's Lemma gives further interesting information if we consider irreducible representations over the real numbers, rather than the complex numbers. Let M be a real irreducible representation of G. Then, when we regard M as a complex representation, one of three things may happen:

- M remains irreducible;
- M splits into two inequivalent irreducible representations;
- M splits into two equivalent irreducible representations.

We say that M has *type* 1, 0 or -1 in the three cases respectively. Now there are just three finite-dimensional division algebras over $\mathbb{R}$, namely $\mathbb{R}$, $\mathbb{C}$, and $\mathbb{H}$ (the non-commutative quaternion algebra): these possibilities for the centraliser algebra correspond to the three types (in the same order).

The *Frobenius–Schur index* of a character χ, written ϵ_χ, is the type of the (unique) real irreducible representation whose complexification includes χ. It can be shown that it is given by the formula

$$\epsilon_\chi = \frac{1}{|G|} \sum_{g \in G} \chi(g^2).$$

Theorem 2.8 *Let G be a permutation group with permutation character $\pi = \sum_{\chi \in \mathrm{Irr}(G)} m_\chi \chi$. Then the number of self-paired orbitals of G is given by*

$$\frac{1}{|G|}\sum_{g\in G}\pi(g^2) = \sum_{\chi\in\mathrm{Irr}(G)} \epsilon_\chi m_\chi.$$

Proof. The above formula for ϵ_χ shows that the two sides of this equation are equal.

Let m be the number of orbits of G on Ω, r the rank, and s the number of self-paired orbitals (so that the number of pairs of non-self-paired orbitals is $(r-s)/2$). Consider also the action of G on the set of 2-subsets of Ω. The permutation character of this action is given by

$$\pi^{\{2\}}(g) = \pi(g)(\pi(g)-1)/2 + c_2(g),$$

where $c_2(g)$ is the number of 2-cycles in the cycle decomposition of g; and the number of orbits in the action is given by $(s-m)+(r-s)/2$ (the two terms counting respectively the non-diagonal self-paired orbitals, and the pairs of non-self-paired orbitals). In addition, we have $\pi(g^2) = \pi(g) + 2c_2(g)$; so

$$\pi^{\{2\}}(g) = \frac{\pi(g)^2 + \pi(g^2)}{2} - \pi(g).$$

Taking inner products of these characters with the principal character, and doing some manipulation, gives the result.

Since the permutation character is the character of a real representation, any irreducible of type -1 must appear in it with even multiplicity. In particular we see that all the orbitals are self-paired if and only if all the irreducibles in π are of type 1 and have multiplicity 1.

More generally, the shape of the centraliser algebra in the last section shows that it is commutative if and only if every constituent of π has multiplicity 1. (The character π is called *multiplicity-free* if this condition holds.) If π is multiplicity-free, then it cannot contain a constituent of type -1; and we see that the numbers of self-paired and non-self-paired orbitals are equal to the numbers of real and non-real constituents of π respectively.

Apply the above result to the group G^*, whose permutation character is indeed multiplicity-free. The orbitals of G^* correspond to the conjugacy classes of G, and an orbital is self-paired if and only if the conjugacy class contains the inverses of its elements. Such a conjugacy class is called *real*. Thus we have:

Theorem 2.9 *The numbers of real and non-real conjugacy classes of a group G are equal to the numbers of real and non-real irreducible characters respectively.*

We can also apply the result to the regular representation of G. In this case, orbitals correspond to elements of G (the orbital corresponding to g consists of all pairs (x, gx) for $x \in G$), and an orbital is self-paired if and only if the corresponding element is equal to its inverse. Since $\rho_G = \sum_{\chi \in \mathrm{Irr}(G)} \chi(1)\chi$, we obtain:

Theorem 2.10 *The number of solutions of $g^2 = 1$ in the group G is given by*

$$\sum_{\chi \in \mathrm{Irr}(G)} \epsilon_\chi \chi(1).$$

2.8 Parker's Lemma

Recently, Richard Parker discovered a generalisation of the Orbit-Counting Lemma, as follows. Let G be a permutation group on Ω, a finite set of cardinality n. Let $\mathcal{C}_k$ denote the set of all k-cycles (in the symmetric group S_n) which occur in the cycle decomposition of some element of G. Note that $\mathcal{C}_1$ consists of all possible 1-cycles – consider the identity – and so is naturally in one-to-one correspondence with Ω. Let $\mathcal{C} = \bigcup_{k=1}^{n} \mathcal{C}_k$. Now G acts on each set $\mathcal{C}_k$ by conjugation: let P_k denote the number of orbits. Furthermore, let $c_k(g)$ be the number of k-cycles in the cycle decomposition of g.

Theorem 2.11 *With the above notation,*

$$P_k = \frac{1}{|G|} \sum_{g \in G} k c_k(g).$$

Proof. Let C be a k-cycle occurring in the element $g \in G$. Let H be its pointwise stabiliser in G. Then the set of elements of G which contain the cycle C is just the coset Hg. The stabiliser of C (in the action of G by conjugation) is $H\langle g \rangle$, and so the size of the orbit of C is $|G : H\langle g \rangle| = |G : H|/k$.

Write out all the permutations in G in an $|G| \times n$ array. Then each of the $|G : H|/k$ cycles in the orbit of C occurs $|Hg| = |H|$ times, covering $|H|k$ cells in the array; so the number of cells covered by cycles in this orbit is

$$|G : H|/k \cdot |H|k = |G|.$$

The total number of cells covered by all the k-cycles in G is

$$\sum_{g \in G} k c_k(g),$$

and so the number of orbits is this number divided by $|G|$.

We noted that C_1 is isomorphic to Ω as a G-space. So the case $k = 1$ of the theorem is the Orbit-Counting Lemma.

We now discuss some consequences. The first corollary is rather surprising:

Corollary 2.12 *The number of orbits of G on $\mathcal{C}$ is equal to n.*

For clearly

$$\sum_{k=1}^{n} kc_k(g) = n$$

for any element g.

The *Parker vector* of G is the n-tuple $P(G) = (P_1, P_2, \ldots, P_n)$. Its components are non-negative integers and sum to n.

The Parker vector is determined by the permutation character, as follows.

Theorem 2.13 *Let G be a permutation group with permutation character π. Then the components of the Parker vector are given by*

$$P_k = \sum_{i|k} \mu(k/i) S_i,$$

where μ is the Möbius function and

$$S_k = \frac{1}{|G|} \sum_{g \in G} \pi(g^k).$$

Proof. We have

$$\pi(g^k) = \sum_{i|k} ic_i(g),$$

since a point is fixed by g^k if and only if it lies in a cycle of length dividing k. Averaging over G gives

$$S_k = \sum_{i|k} P_i.$$

The result follows by Möbius inversion (see Section 2.11).

The genesis of the Parker vector is interesting. Suppose that we have an unknown permutation group G. Imagine that we can choose random elements of G and determine only their cycle decomposition, not their precise action on Ω. Then from the formula we can calculate an approximation to the Parker vector of G. Since it is an integer vector, after a while we can say with confidence that we know the exact Parker vector.

Parker pointed out that this situation arises in the computation of Galois groups over the rationals (or, which amounts to the same thing, over the integers). Let f be a polynomial with integer coefficients. For any prime p, let f_p denote the reduction of f mod p. Factorise f_p over the integers mod p. The degrees of the irreducible factors are the cycle lengths of the ***Frobenius automorphism*** $x \mapsto x^p$, acting on the roots of f_p. (This is because the Frobenius automorphism generates the Galois group of f_p; its cycles permute transitively the roots of the irreducible factors of f_p.) Now, if p doesn't divide the *discriminant* of f (this involves avoiding finitely many 'bad' primes), then the Frobenius automorphism lifts to an element of the Galois group G of f over $\mathbb{Q}$ with the same cycle structure. Of course there is no way to match up the actions of the Frobenius automorphisms for different primes without solving the polynomial completely. Nevertheless, we might hope that the lifted Frobenius maps for 'random' large primes will be random elements of G. So we can (with this assumption) compute the Parker vector of G.

This motivates Parker's main question:

> What does the Parker vector of a permutation group G tell us about G? In particular, which groups, or classes of groups, are determined by $P(G)$?

Not surprisingly, not every permutation group is determined by its Parker vector. However, some groups are. Also, as we saw, P_1 is the number of orbits of G, and $P_1 + P_2$ is the number of self-paired orbitals. So groups with $P_1 = P_2 = 1$ include all 2-transitive groups; these have been determined using the Classification of Finite Simple Groups, as we will see in Chapter 4. We can also obtain a classification at the other extreme. A regular group G has $P_2 = 1$ if and only if it contains a unique involution z (see Exercise 2.11). Then $Z = \langle z \rangle \trianglelefteq G$. Let $\overline{G} = G/Z$. The Sylow 2-subgroup P of G then also contains a unique involution. A 2-group with this property must be cyclic or a generalised quaternion group. Then $\overline{P} = P/Z$ is cyclic or dihedral. If $\overline{P}$ is cyclic, Burnside's Transfer Theorem shows that $\overline{G}$ has a *normal 2-complement*, a normal subgroup $\overline{N}$ with $\overline{P} \cap \overline{N} = \{1\}$ and $\overline{PN} = \overline{G}$ (so that $\overline{G}/\overline{N} \cong \overline{P}$). If $\overline{P}$ is dihedral, the *Gorenstein–Walter Theorem* shows that there is a normal subgroup $\overline{N}$ of $\overline{G}$ of odd order such that $\overline{G}/\overline{N}$ is isomorphic to a subgroup of $\mathrm{P\Gamma L}(2,q)$ containing $\mathrm{PSL}(2,q)$ for odd q, or to A_7, or to $\overline{P}$. (In the first case, not all subgroups of $\mathrm{P\Gamma L}(2,q)$ containing $\mathrm{PSL}(2,q)$ have dihedral Sylow subgroups.)

In principle, we could now examine these candidates for $\overline{G}$ to determine which are the quotients of a group G by a unique involution. In fact, a cohomological argument of George Glauberman shows that this is not necessary:

Theorem 2.14 *Let $\overline{G}$ be a group with cyclic or dihedral Sylow 2-subgroups. Then there is a unique group G (up to isomorphism) having a unique involution z such that $G/\langle z\rangle \cong \overline{G}$.*

The proof is given in [9].

Block's Lemma asserts that, if G acts on two sets with Parker vectors $P^{(1)}$ and $P^{(2)}$, and the permutation character of the first action is contained in that of the second, then $P_1^{(1)} \leq P_1^{(2)}$. In the lectures, I posed the problem of deciding whether, under these hypotheses, it is true that $P^{(1)}$ is *dominated by* $P^{(2)}$ (in the sense that the inequality holds for every component). This and several variants were refuted by Jürgen Müller, who found a series of examples. This prompted a different analysis of the problem as follows.

Let χ be any character of a finite group G. We can define the *Parker vector* of the character χ by the equations

$$S_k(\chi) = \frac{1}{|G|}\sum_{g\in G}\chi(g^k),$$

$$P_k(\chi) = \sum_{i|k}\mu(k/i)S_i(\chi).$$

If χ is an irreducible character, then $S_2(\chi)$ is its Frobenius–Schur index ϵ_χ. For this reason, the values $S_k(\chi)$ are called *generalised Frobenius–Schur indices* of the irreducible character χ.

From Theorem 2.13, we see that, if χ is a permutation character, then this is the usual Parker vector. In general, the following is true:

Theorem 2.15 *(a) $P_k(\chi) \in \mathbb{Z}$ for all $k > 0$.*

(b) $P_k(\chi) = 0$ if k does not divide $|G|$.

(c) $\sum_{k\geq 1} P_k(\chi) = \chi(1)$.

Note that the sum in (c) is finite because of (b).

Proof. (a) It suffices to prove that $S_k(\chi) \in \mathbb{Z}$ for all $k > 0$. Now consider a particular $g \in G$. The matrix $M(g)$ representing g in the representation with character χ is diagonalisable: for, if $g^m = 1$, then the minimal polynomial of $M(g)$ divides $x^m - 1$, which has distinct roots over $\mathbb{C}$. So $\chi(g) = \sum \lambda_i^k = p_k(\lambda_1, \ldots)$, where $\lambda_1, \ldots$ are the eigenvalues, and p_k is the kth *power sum function*.

It follows from *Newton's Theorem* on symmetric functions that p_k can be written as a polynomial with integer coefficients in the *elementary symmetric functions* $e_1, e_2, \ldots$, where e_i is the sum of all products of i distinct arguments. Now observe that

- $e_i(\lambda_1, \ldots) = \mathrm{Trace} \bigwedge^i M(g)$, where $\bigwedge^i M$ is the matrix representation on the ith exterior power of the original vector space; and
- $\mathrm{Trace}\, M_1(g)\, \mathrm{Trace}\, M_2(g) = \mathrm{Trace}(M_1 \otimes M_2)(g)$, where $M_1 \otimes M_2$ is the representation on the tensor product of the spaces affording M_1 and M_2.

From this, we conclude that $\chi(g^k)$ is a *virtual character* of G, a sum of characters with integer coefficients. Then $S_k(\chi)$ is the inner product of this virtual character with the principal character, and so is an integer, as required.

(b) First we note that $S_k(\chi) = S_d(\chi)$ where d is the greatest common divisor of k and $n = |G|$. This holds because the kth powers and the dth powers run through the same elements of G with the same multiplicities. (The map $g \mapsto g^{k/d}$ is a bijection.)

Suppose that k does not divide n. Then there is a prime p dividing k to a higher power than n. The only non-zero terms in the sum defining P_k are those for which k/i is squarefree. The values of i can be divided into pairs i, ip, where k/ip is not divisible by p. Now $(i, n) = (ip, n)$, so $S_i = S_{ip}$; and $\mu(k/i) = -\mu(k/ip)$. So paired terms cancel, and the sum is zero.

(c) According to (b), the sum in the theorem is just

$$\sum_{k|n} P_k(\chi),$$

which is equal to $S_n(\chi)$, by Möbius inversion; and since $g^n = 1$ for all $g \in G$, we have $S_n(\chi) = \chi(1)$.

Problem. For which characters χ are the $P_k(\chi)$ all non-negative? (Note that the Parker vector of a permutation action of a group is dominated by the Parker vector of another action if and only if the difference of the characters has all P_k non-negative.) It seems, empirically, that there is a tendency for the values to be non-negative!

2.9 Characters of abelian groups

Let G be an abelian group of order n. Then G has n conjugacy classes, and so n irreducible characters. Since the sum of squares of their degrees is n, each character has degree 1. Thus, each character is simply a homomorphism from G to the multiplicative group of non-zero complex numbers.

It follows that the product of any two characters is a character. So the set G^* of irreducible characters of G is itself a group: the identity element is the principal character, and the inverse of a character is its complex conjugate.

To determine the structure of G^*, we invoke the *Fundamental Theorem of Abelian Groups*, which asserts that G can be written as a direct product of cyclic groups. Now, if $A = \langle a \rangle$ is a cyclic group of order n, and $\omega = e^{2\pi i/n}$ a primitive nth root of unity, then for $0 \leq i \leq n-1$, there is a character χ_i of G, given by $\chi_i(a^j) = \omega^{ij}$, and these are all the characters of G. We see that $\chi_i = (\chi_1)^i$, so that G^* is also cyclic, generated by χ_1.

Now let $G = C_1 \times \cdots \times C_r$ be an arbitrary finite abelian group. We have seen that $C_i^* \cong C_i$, and we also observed that any character of a direct product is a product of characters of the factors. It follows that

$$G^* = C_1^* \times \cdots \times C_r^* \cong C_1 \times \cdots \times C_r = G.$$

However, the isomorphism between G and its dual is not canonical: it depends on the direct sum decomposition of G and on the choice of primitive roots of unity. There is, however, a canonical isomorphism between G and its second dual G^{**}: for each $g \in G$, we define a character Ξ_g of G^* by the rule

$$\Xi_g(\chi) = \chi(g).$$

The map $g \mapsto \Xi_g$ is a monomorphism; since $|G| = |G^{**}|$, it is an isomorphism.

Theorem 2.16 *The characters of an abelian group G form a group G^* isomorphic to G. For any subgroup H of G, the characters of G whose restriction to H is the trivial character form a subgroup $H^\dagger$ of H; and $|H^\dagger| = |G : H|$.*

Proof. Only the last sentence remains to be proved. It is clear that restriction to H is a homomorphism from G^* to H^*, whose kernel is $H^\dagger$; so $H^\dagger$ is a subgroup of G, with order at least $|G^*|/|H^*| = |G : H|$. On the other hand, any character of G with H in its kernel induces in a natural way a character of G/H; so $|H^\dagger| \leq |G/H|$. So equality holds.

Any function f from G to $\mathbb{C}$ can be written as a linear combination of characters, since the characters span the space of class functions. So

$$f(g) = \sum_{\chi \in G^*} f^*(\chi)\chi(g),$$

where f^* is a function from G^* to $\mathbb{C}$. The function f^* is called the *Fourier transform* of f. From the orthogonality relations, we have the *inversion formula*

$$f^*(\chi) = \frac{1}{|G|} \sum_{g \in G} f(g)\overline{\chi(g)}.$$

(See Exercise 2.21.)

Remark. The duality of abelian groups and their character groups can be developed in a much more general context, that of locally compact abelian groups. In particular, the dual of the abelian group $G = \mathbb{R}/\mathbb{Z}$ is the discrete group $G^* = \mathbb{Z}$. A function on G is just a periodic function on $\mathbb{R}$, and its Fourier transform is the function on G^* giving the usual Fourier coefficients. We do not pursue this here. However, the duality for finite abelian groups has many applications in coding theory, statistics, etc.

2.10 Characters of the symmetric group

The character theory of the symmetric group is a vast subject, closely connected with symmetric functions and parts of combinatorics. This section gives a very brief outline without proofs. For more details see Sagan [158], Macdonald [123] or Fulton [78].

It is well known that two elements of S_n are conjugate if and only if they have the same *cycle structure*: that is, the lists (or multisets) of cycle lengths of the two permutations are the same. So the number of conjugacy classes in S_n is the *partition number* $p(n)$, the number of partitions of n. We represent a partition λ of n as

$$n = n_1 + \cdots + n_k, \qquad n_1 \geq \cdots \geq n_k > 0.$$

We sometimes denote this partition by $(n_1, \ldots, n_k)$.

From the general theory, we know that the number of irreducible characters of S_n is also equal to $p(n)$. How can we match characters to partitions?

Let λ be a partition of n as above. Let Ω^λ be the set consisting of all k-tuples $(X_1, \ldots, X_k)$ where $X_1, \ldots, X_k$ are subsets of Ω satisfying $|X_i| = n_i$, $X_i \cap X_j = \emptyset$ for $i \neq j$. (It follows that $X_1 \cup \cdots \cup X_k = \Omega$. However, Ω^λ is not the set of partitions of Ω of shape λ: why not?) Then S_n acts on Ω^λ in a natural way; the stabiliser of an element of Ω^λ is a *Young subgroup* S_λ of S_n, and is isomorphic to $S_{n_1} \times \cdots \times S_{n_k}$.

Let π^λ be the permutation character of G on Ω^λ.

With every partition λ is associated a *dual partition* λ^*, of the form

$$n = l_1 + \cdots + l_h, \qquad l_1 \geq \cdots \geq l_h > 0,$$

where l_h is the number of parts n_i of λ which satisfy $n_i \geq h$. (So $l_1 = k$.) We introduce the *diagram* (or *Young diagram*) $D(\lambda)$ of a partition λ: take n boxes and arrange them in rows containing $n_1, \ldots, n_k$ boxes, left-aligned. The diagram of the partition $(3,2)$ is as shown in Figure 2.1. Now the diagram of the dual partition is obtained simply by transposing $D(\lambda)$ (so that $(3,2)^* = (2,2,1)$).

There is a natural partial order on partitions, defined as follows: if $\lambda = (n_1, \ldots, n_k)$ and $\mu = (m_1, \ldots, m_r)$, then $\lambda \leq \mu$ if and only if $n_1 + \cdots + n_i \geq m_1 + \cdots + m_i$ for $i = 1, \ldots, r$ (adjoining zero parts if necessary). The first part in the order is (n) and the last is $(1, 1, \ldots, 1)$. It can be shown that duality reverses the order: $\lambda \leq \mu$ if and only if $\mu^* \leq \lambda^*$.

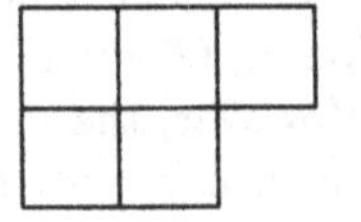

Figure 2.1: $D((3,2))$

Finally, there is a *sign character* sign on S_n, taking the value $+1$ or -1 on G according as g is an even or an odd permutation. Let $\tau^\lambda = \pi^\lambda \cdot \mathrm{sign}$.

Now we can state three facts about the irreducible characters of S_n.

Theorem 2.17 *For each partition λ of n, there is an irreducible character χ^λ, satisfying*

(a) χ^λ is a constituent of π^λ but not of π^μ for any $\mu < \lambda$,

(b) χ^λ is the unique common constituent (with multiplicity 1) of π^λ and τ^{λ^*},

(c) $\chi^\lambda \cdot \mathrm{sign} = \chi^{\lambda^*}$.

Since we have an irreducible character for each partition, and these characters are all distinct, we have all the characters! Note that they are all integer-valued (and of Frobenius–Schur index 1), since they are linear combinations of permutation characters. This is because, by (a), the matrix expressing π^λ in terms of χ^μ is lower triangular, with ones on the diagonal.

Example. In working out this example, we use the fact that the inner product of two permutation characters is equal to the number of orbits of G on the cartesian product of the two sets. Now $\Omega^{(n)}$ has a single point and so $\pi^{(n)}$ is the principal character; and $\Omega^{(n-1,1)}$ is isomorphic to Ω (if $n > 2$), so $\pi^{(n-1,1)}$ is the usual permutation character. Similarly, if $n > 4$, then $\pi^{(n-2,2)}$ and $\pi^{(n-2,1,1)}$ are the permutation characters of S_n on unordered and ordered pairs of distinct points respectively. The various inner products can be worked out: for example,

$$\begin{aligned}
\langle \pi^{(n-1,1)}, \pi^{(n-1,1)} \rangle &= 2,\\
\langle \pi^{(n-1,1)}, \pi^{(n-2,2)} \rangle &= 2,\\
\langle \pi^{(n-1,1)}, \pi^{(n-2,1,1)} \rangle &= 3,\\
\langle \pi^{(n-2,2)}, \pi^{(n-2,2)} \rangle &= 3,\\
\langle \pi^{(n-2,2)}, \pi^{(n-2,1,1)} \rangle &= 4,\\
\langle \pi^{(n-2,1,1)}, \pi^{(n-2,1,1)} \rangle &= 6.
\end{aligned}$$

For example, consider $\langle \pi^{(n-2,2)}, \pi^{(n-2,1,1)} \rangle$. This is the number of orbits of S_n on configurations $(\{\alpha,\beta\},(\gamma,\delta))$ with $\alpha \neq \beta$ and $\gamma \neq \delta$. There are four such orbits, corresponding to the cases where $\{\alpha,\beta\} \cap \{\gamma,\delta\}$ is $\{\gamma,\delta\}$, $\{\gamma\}$, $\{\delta\}$, or $\emptyset$.

From this, we easily obtain

$$\begin{aligned}
\pi^{(n)} &= \chi^{(n)} = 1, \\
\pi^{(n-1,1)} &= 1 + \chi^{(n-1,1)}, \\
\pi^{(n-2,2)} &= 1 + \chi^{(n-1,1)} + \chi^{(n-2,2)}, \\
\pi^{(n-2,1,1)} &= 1 + 2\chi^{(n-1,1)} + \chi^{(n-2,2)} + \chi^{(n-2,1,1)}.
\end{aligned}$$

These are found recursively. For example, let $\pi = \pi^{n-2,1,1}$. First, $\langle \pi, \pi^{(n)} \rangle = 1$ shows that π contains $\chi^{(n)}$ with multiplicity 1. Then $\langle \pi, \pi^{(n-1,1)} \rangle = 3$ shows that $\chi^{(n-1,1)}$ occurs with multiplicity 2. Then $\langle \pi, \pi^{(n-2,2)} \rangle = 4$ shows that $\chi^{(n-2,2)}$ has multiplicity 1; and finally $\langle \pi, \pi \rangle = 6$ shows that there is just one more irreducible, which must be $\chi^{(n-2,1,1)}$.

There are important connections with *symmetric functions*. For example, if p_λ is the *power sum function* defined by $p_n = \sum x_i^n$ and $p_{(n_1,\ldots,n_k)} = p_{n_1} \cdots p_{n_k}$, and s_λ is the *Schur function*, then

$$s_\lambda = \sum a_{\lambda\mu} p_\mu,$$

where the matrix $(a_{\lambda\mu})$ is the character table of S_n.

We now give two remarkable formulae for the degree f^λ of the character χ^λ.

A *Young tableau* of shape λ is an arrangement of the numbers $1, 2, \ldots, n$ into the cells of the diagram $D(\lambda)$ in such a way that the rows and columns of the array are increasing. (Strictly, this is a 'standard Young tableau', but we will not consider any non-standard tableaux.)

The *hook* $H(i,j)$ at cell (i,j) in the diagram $D(\lambda)$ is the set of cells to the right or below the cell (i,j) (including this cell itself). The *hook length* $h(i,j)$ is the number of cells in the hook $H(i,j)$.

Theorem 2.18 *Let f^λ be the degree of the character χ^λ.*

(a) f^λ is the number of Young tableaux of shape λ.

(b)

$$f^\lambda = \frac{n!}{\prod_{(i,j)\in D(\lambda)} h(i,j)}.$$

Part (b) is called the *hook length formula*. A proof is given in Sagan [158].

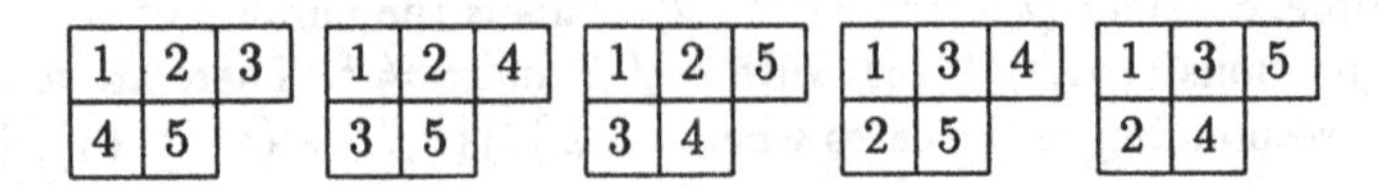

Figure 2.2: Young tableaux

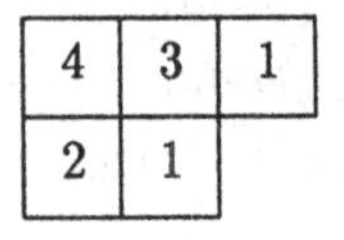

Figure 2.3: Hook lengths

Example. For the partition (3,2) of 5, Figure 2.2 gives the standard tableaux and Figure 2.3 the hook lengths.

The hook length formula is verified by the calculation

$$\frac{5!}{4\cdot 3\cdot 1\cdot 2\cdot 1} = 5.$$

Again, from Theorems 2.7 and 2.10, since all characters have Frobenius–Schur index 1, we have

$$\begin{aligned} n! &= \sum_{\lambda}(f^\lambda)^2, \\ |\{g \in S_n : g^2 = 1\}| &= \sum_{\lambda} f^\lambda. \end{aligned}$$

These equations are explained combinatorially by the *Robinson–Schensted correspondence*, a bijection between S_n and the set of pairs of tableaux of the same shape, having the property that, if g corresponds to (S,T), then g^{-1} corresponds to (T,S) (so that $S = T$ if and only if $g^2 = 1$).

2.11 Appendix: Möbius inversion

The *Möbius function* is the function μ on the positive integers given by

$$\mu(n) = \begin{cases} (-1)^k & \text{if } n = p_1 \cdots p_k, \text{ where} \\ & \quad p_1, \ldots, p_k \text{ are distinct primes,} \\ 0 & \text{if } n \text{ is not squarefree.} \end{cases}$$

Its important property is:

Theorem 2.19 *Let F and G be functions on the positive integers. If*

$$G(n) = \sum_{d|n} F(d),$$

then

$$F(n) = \sum_{d|n} \mu(n/d)G(d),$$

and conversely.

The proof depends on the property that

$$\sum_{d|n} \mu(d) = \begin{cases} 1 & \text{if } n = 1, \\ 0 & \text{otherwise.} \end{cases}$$

This is clear if $n = 1$. If $n > 1$, then choose a prime p dividing n; partition the squarefree divisors of n into pairs d, dp, where d is not divisible by p. Then $\mu(dp) = -\mu(d)$, so the contributions cancel. The non-squarefree divisors give zero by definition.

Now let $G(n) = \sum_{d|n} F(d)$. Then

$$\sum_{d|n} \mu(n/d)G(d) = \sum_{d|n} \sum_{e|d} \mu(n/d)F(d) = \sum_{e|n} \left(\sum_{m|n/e} \mu(m) \right) F(e) = F(n),$$

on setting $m = n/d$. The converse is proved similarly.

Remark. This is the tip of a very large iceberg. For any partially ordered set in which all intervals are finite, there is an integer-valued Möbius function of two variables, $\mu(x, y)$, giving an inversion relation: if

$$G(x, y) = \sum_{x \leq z \leq y} F(x, z),$$

then

$$F(x, y) = \sum_{x \leq z \leq y} G(x, z)\mu(z, y),$$

and conversely. If we take the positive integers ordered by divisibility, then the function $\mu(1, n)$ is the 'classical' Möbius function above.

2.12 Exercises

2.1. Prove that two permutation representations of a cyclic group which have the same permutation character are isomorphic. Give an example to show that this is false for arbitrary groups.

2.2. Let G be a transitive permutation group of degree $n > 1$.

(a) Prove that G contains a fixed-point-free element (one with no fixed points).

(b) Prove that the number of fixed-point-free elements in G is at least $|G|/n$, with equality if and only if G is sharply 2-transitive.

2.3. Prove that a subgroup of index less than n in a 2-transitive group of degree n is transitive.

2.4. Let G be a finite group. Prove that the number of conjugacy classes in G is given by

$$\frac{1}{|G|}\sum_{g\in G}|C_G(g)|,$$

where $C_G(g) = \{x \in G : gx = xg\}$ is the *centraliser* of g.

2.5. Consider the following Markov chain on the finite group G: a transition consists in moving from the element g to a randomly chosen element that commutes with g (all such elements equally likely). Prove that, in the limiting distribution, all conjugacy classes in G are equally likely. What is the connection with Jerrum's Markov chain? (I am grateful to Leonard Soicher for this exercise.)

2.6. Let G act transitively on Ω_1 and Ω_2, with ranks r_1 and r_2 respectively. Suppose that the permutation character of G on Ω_1 is contained in the permutation character on Ω_2. Prove that $r_1 \leq r_2$.

2.7. Let G act on Ω. A subset S of G is said to be *uniformly transitive* if, for all $\alpha, \beta \in \Omega$, the number of elements $g \in S$ with $\alpha g = \beta$ is constant. It is *sharply transitive* if this number is 1. Moreover, we say that S is *uniformly/sharply k-transitive* on Ω if it is uniformly/sharply transitive on the set of ordered k-tuples of distinct elements of Ω.

(a) Let S be uniformly transitive on Ω. Show that $P(S) = \lambda J$, where P is the permutation representation, J the all-1 matrix, and $P(S) = \sum_{g\in S} P(g)$. Deduce that $M(S) = O$, where M is any non-principal irreducible representation which is a constituent of P. Show that, conversely, if this condition holds, then S is uniformly transitive.

(b) Suppose that G acts on Γ and Δ. Suppose that every irreducible constituent of the permutation character on Δ occurs as a constituent of the permutation character on Γ. Prove that a subset which is uniformly transitive on Γ is uniformly transitive on Δ. Hence prove that if there is a subset sharply transitive on Γ, then $|\Delta|$ divides $|\Gamma|$.

(c) Taking $G = \mathrm{PGL}(n, q)$ with $n > 2$, Γ the set of ordered pairs of distinct points of the projective space, and Δ the set of *antiflags* (non-incident point–hyperplane pairs), the condition of (b) is satisfied. Deduce that G contains no subset which is sharply 2-transitive on points.

[*Note:* The existence of a sharply 2-transitive set of permutations is equivalent to that of a projective plane. The above argument is due to M. O'Nan [144].]

2.8. In how many ways can the faces of a cube be coloured with m colours, if colourings related by a rotation of the cube are regarded as being the same?

2.9. Let G and H be finite groups. Using the method outlined in Section 2.6, prove that, for any $\phi \in \mathrm{Irr}(G)$ and $\psi \in \mathrm{Irr}(H)$, the function $[\phi, \psi]$ on $G \times H$ given by

$$[\phi, \psi](g, h) = \phi(g)\psi(h)$$

is an irreducible character. Prove that every irreducible character arises in this way. Show also that

$$\epsilon_{[\phi,\psi]} = \epsilon_\phi \epsilon_\psi.$$

2.10. Find the character table of the symmetric group S_4. Show the identification of the irreducible characters with partitions. Verify the tableau and hook-length formulae for the character degrees. Now try the same question for S_5.

2.11. Show that the ith component P_i of the Parker vector of a regular group G is equal to the number of elements of order i in G.

2.12. Let G_1 and G_2 be permutation groups on Ω_1 and Ω_2 respectively, with Parker vectors $P(G_1)$ and $P(G_2)$ respectively, which we regard as infinite sequences with all but finitely many terms zero. In what follows, $G_1 \times G_2$ acts on the disjoint union of Ω_1 and Ω_2, and $G_1 \mathrm{\,Wr\,} G_2$ acts on $\Omega_1 \times \Omega_2$.

(a) Prove that $P(G_1 \times G_2) = P(G_1) + P(G_2)$.

(b) Prove that $P(G_1 \mathrm{\,Wr\,} G_2) = P(G_2 \mathrm{\,Wr\,} G_1) = P(G_1) \circ P(G_2)$, where $x \circ y$ is given by $(x \circ y)_k = \sum_{i|k} x_i y_{k/i}$.

(I am grateful to Julian Gilbey for this observation.)

2.13. Show that, if χ is a character of G, then $P_k(\chi) = 0$ if k does not divide the exponent of G. (I am grateful to Jürgen Müller for this observation.)

2.14. Let G be a finite transitive permutation group. Let H be the subgroup of G generated by all the elements of G which have no fixed points. Show that H is a normal subgroup of G, that H is transitive, and that H contains every element $g \in G$ with $\pi(g) \neq 1$. Give an example where $H \neq G$. (This result is due to Cohen and Zantema [52].)

2.15. Let $\lambda = (n_1, n_2, \ldots, n_r)$ be a partition of n, with $n_1 \geq n_2 \geq \cdots \geq n_r$. Let χ^λ be the corresponding character of S_n. We define the *dimension* of χ_λ to be $n - n_1$. Prove that, for fixed d and $n > 2d$,

(a) the number of characters of dimension d is $p(d)$;

(b) the degree of a character χ^λ of dimension d is a polynomial in n with degree d and leading coefficient $f_{\lambda_1}/d!$, where λ_1 is the partition of d obtained by removing the part $n - d$ from λ, and f_{λ_1} its degree.

2.16. Let $n_2(G)$ denote the number of involutions in G.

(a) Prove that $n_2(G) \leq \sqrt{r(|G| - 1)}$, where r is the number of real conjugacy classes of G different from $\{1\}$.

(b) Prove that, if all irreducible characters of G have Frobenius–Schur index 1, then $n_2(G) \geq \sqrt{|G| - 1}$.

2.17. The aim of this exercise is to prove a theorem of Sims [167]: If Γ is an orbital of a transitive permutation group G, then corresponding *suborbits* are the G_α-orbit $\Gamma(\alpha) = \{\beta : (\alpha, \beta) \in \Gamma\}$, for each $\alpha \in \Omega$; and the *subconstituent* of G is the permutation group $G_\alpha^{\Gamma(\alpha)}$ induced on $\Gamma(\alpha)$ by G_α. Sims showed:

Theorem 2.20 *If G is a finite transitive permutation group with a suborbit $\Gamma(\alpha)$ of cardinality greater than 1, then the paired subconstituents $G_\alpha^{\Gamma(\alpha)}$ and $G_\alpha^{\Gamma^*(\alpha)}$ have a non-trivial common homomorphic image.*

(a) Let G act transitively on sets Γ and Δ, and suppose that G^Γ and G^Δ have no non-trivial homomorphic image in common. Show that $G^{\Gamma \cup \Delta} = G^\Gamma \times G^\Delta$, and deduce that $G_\gamma^\Delta = G^\Delta$ and $G_\delta^\Gamma = G^\Gamma$ for $\gamma \in \Gamma$ and $\delta \in \Delta$.

(b) Suppose that G is transitive on Ω and the paired subconstituents $G_\alpha^{\Gamma(\alpha)}$ and $G_\alpha^{\Gamma^*(\alpha)}$ have no non-trivial common homomorphic image. Define an *n-path* to be a sequence $(\alpha_0, \ldots, \alpha_n)$ for which $(\alpha_{i-1}, \alpha_i) \in \Gamma$ for $i = 1, \ldots, n$. Prove the following two statements by induction on n:

(i) G acts transitively on n-paths;

(ii) if H is the stabiliser of an $(n-1)$-path $(\alpha_1, \ldots, \alpha_n)$, then $H^{\Gamma(\alpha_n)} = G_{\alpha_n}^{\Gamma(\alpha_n)}$ and $H^{\Gamma^*(\alpha_1)} = G_{\alpha_1}^{\Gamma^*(\alpha_1)}$.

(c) Hence show that, if Ω is finite, then $|\Gamma(\alpha)| = 1$.

2.18. The following variation on Sims' result is due to Cameron [30, I].

(a) Show that, if Γ and Δ are finite sets of the same size, and G acts on them such that G^Γ is 2-transitive and G^Δ is not, then G_δ^Γ is 2-transitive and G_γ^Δ is not, for $\gamma \in \Gamma$, $\delta \in \Delta$. [*Hint:* Consider the permutation characters.]

(b) Hence show, using the method of the preceding exercise, that if a subconstituent of a finite transitive permutation group is 2-transitive, then so is the paired subconstituent.

2.19. Show that any two transitive permutation groups occur as paired subconstituents in some infinite transitive permutation group. [*Hint:* The HNN-construction shows that, if A and B are isomorphic subgroups of a group G, then there is a transitive permutation group $G' = \langle G, t\rangle$ in which G is a stabiliser G'_α and A and B are 2-point stabilisers $G_{\alpha\beta}$ and $G_{\alpha\gamma}$, where $(\alpha, \beta) = (\gamma, \alpha)t$.]

2.20. Show that a non-trivial irreducible constituent of the permutation character of a primitive permutation group is faithful.

2.21. Prove the Fourier inversion formula.

2.22. The Mathieu group M_{23} is a 4-transitive permutation group on the set $\Omega = \{1, \ldots, 23\}$, and preserves a set $\mathcal{B}$ of 253 subsets of Ω, each of size 7. *From this information alone*, show that M_{23} acts transitively on $\mathcal{B}$ with rank 3.

2.23. Let G be a group with finitely many subgroups of index n for all n. Let $c_n(G)$ be the number of conjugacy classes of subgroups of index n in G, and $d_n(G)$ the number of G-spaces with n points (up to isomorphism). Prove that

$$\sum_{n=0}^{\infty} d_n(G)t^n = \exp\left(\sum_{k=1}^{\infty} \frac{c(t^k)}{k}\right),$$

where

$$c(t) = \sum_{n=1}^{\infty} c_n t^n.$$

(If G is a finite group, the function $c(t)$ is a polynomial, and so the two expressions are complex analytic functions. If G is infinite, $c(t)$ is a power series which may not converge, and the expressions are formal power series with no analytic interpretation.)

2.24. (a) Let G be a group, H a subgroup of order n. Prove that the number of homomorphisms $\phi : G \to S_n$ in which the inverse image of S_{n-1} (the stabiliser of the point n) is H is equal to $(n-1)!$.

(b) Hence show that, if G is a group for which the number $s_n(G)$ of subgroups of index n is finite, then the number of homomorphisms $\phi : G \to S_n$ such that the image of G under ϕ is a transitive subgroup of S_n is $(n-1)!s_n(G)$.

(c) Hence show the following. Let G be a group with finitely many subgroups of index n for all n. Let $s_n(G)$ be the number of subgroups of index n in G, and $h_n(G)$ the number of homomorphisms from G to S_n. Then, setting $h_0(G) = 1$, we have

$$\sum_{n=0}^{\infty} \frac{h_n(G)}{n!} t^n = \exp\left(\sum_{n=1}^{\infty} \frac{s_n(G)}{n} t^n\right).$$

2.25. Prove that a finitely generated group has only finitely many subgroups of index n for all n. Interpret the equations of the last two questions in the case where G is the infinite cyclic group.

CHAPTER 3

Coherent configurations

3.1 Introduction

Let G be a permutation group on Ω. We saw in Chapter 1 that any *orbital* of G (orbit of G on Ω^2) can be represented by an *orbital graph*. In this section, we consider a more complicated combinatorial object called a *coherent configuration* which describes all the orbitals simultaneously. Coherent configurations were introduced by Donald Higman in [90] and a number of subsequent papers, but the idea goes back much further. A special case was introduced into statistics by Bose and Nair [24] in 1939 under the name 'association scheme'. The roots in the centraliser algebra of a group can be traced to the work of Brauer in the 1930s. The general concept was also defined by Weisfeiler and Leman [179] at about the same time as Higman, and studied extensively in the former Soviet Union under the name 'cellular algebra' (see [71]).

Let the permutation group G have m orbits on Ω, and have rank r. An orbital is a subset of Ω, that is, a set of ordered pairs. It is customary to regard such a set of pairs as a binary relation on Ω. Accordingly, we call the orbitals $R_1, \ldots, R_r$ and regard them as relations on Ω. If $\Omega_1, \ldots, \Omega_m$ are the orbits, we choose the notation so that, for $1 \leq i \leq m$,

$$R_i = \{(\alpha, \alpha) : \alpha \in \Omega_i\}$$

is the diagonal of Ω_i. Let $\mathcal{R} = \{R_1, \ldots, R_r\}$.

The relations R_i have the following properties:

(CC1) $\mathcal{R}$ is a partition of Ω^2.

(CC2) There is a subset of $\mathcal{R}$ which is a partition of the diagonal of Ω^2.

(CC3) $\mathcal{R}$ is closed under transposition (where the transpose of R is $R^* = \{(\beta, \alpha) : (\alpha, \beta) \in R\}$).

(CC4) Given $i, j, k \in \{1, \ldots, r\}$ and $(\alpha, \beta) \in R_k$, the number of points $\gamma \in \Omega$ with $(\alpha, \gamma) \in R_i$ and $(\gamma, \beta) \in R_j$ depends only on i, j, k, not on the choice of $(\alpha, \beta) \in R_k$.

The subset referred to in (CC2) is $\{R_1, \ldots, R_m\}$. The conditions (CC1) and (CC3) are clear; and (CC4) holds because R_k is a G-orbit. (If $(\alpha, \beta), (\alpha', \beta') \in R_k$ and g maps (α, β) to (α', β'), then g takes the set

$$\{\gamma \in \Omega : (\alpha, \gamma) \in R_i, (\gamma, \beta) \in R_j\}$$

to the set

$$\{\gamma \in \Omega : (\alpha', \gamma) \in R_i, (\gamma, \beta') \in R_j\}$$

and so these sets have the same cardinality.) We denote the number of points referred to in (CC4) by p_{ij}^k.

Definition. (a) A *coherent configuration* is a pair $(\Omega, \mathcal{R})$, where Ω is a set, and $\mathcal{R}$ a set of binary relations on Ω satisfying conditions (CC1)–(CC4).

(b) If G is a permutation group on Ω, and $\mathcal{R}$ the set of orbitals of G, then $(\Omega, \mathcal{R})$ is the *coherent configuration of* G (or *associated with* G).

The numbers p_{ij}^k were called the *intersection numbers* of the configuration by Higman. The reason is as follows. For $\alpha \in \Omega$ and $1 \leq i \leq r$, we set

$$R_i(\alpha) = \{\beta \in \Omega : (\alpha, \beta) \in R_i\}.$$

Then p_{ij}^k is the cardinality of the intersection $R_i(\alpha) \cap R_j^*(\beta)$, for $(\alpha, \beta) \in R_k$.

Not every coherent configuration is associated with a group. Define an *automorphism* of $(\Omega, \mathcal{R})$ to be a permutation of Ω which preserves each relation in R. Now, if a coherent configuration $(\Omega, \mathcal{R})$ is associated with a group G, then G is a subgroup of the automorphism group of $(\Omega, \mathcal{R})$, and acts transitively on each relation in $\mathcal{R}$. It is known that many coherent configurations have no non-trivial automorphisms.

On the other hand, it can happen that a coherent configuration 'is' a group. Suppose that, for all $i \in \{1, \ldots, r\}$ and all $\alpha \in \Omega$, there exists a unique $\beta \in \Omega$ such that $(\alpha, \beta) \in R_i$. Now set $R_i \circ R_j = R_k$ whenever $p_{ij}^k > 0$ (there is a unique value of k for which this holds, given any i, j). Then it is easily seen that we obtain a group G, and the coherent configuration is associated with the regular representation of G. Any group arises in this way, on taking the coherent configuration associated with its regular representation. Thus, coherent configurations form a generalisation of groups!

The definition can be phrased in the language of matrices. Given a relation R on Ω, we define its *basis matrix* $B(R)$ to be the matrix with rows and columns indexed by Ω, with (α, β) entry 1 if $(\alpha, \beta) \in R$, 0 otherwise. In other words, it is the characteristic function of R. In the sequel, I denotes the identity matrix, and J the matrix with every entry equal to 1, of the appropriate size. The following result is simply the translation of the definition into matrix terms.

Theorem 3.1 *Let $\mathcal{A} = \{A_1, \ldots, A_r\}$ be a set of zero–one matrices of the same size. Then $\mathcal{A}$ is the set of basis matrices of a coherent configuration if and only if the following conditions are satisfied:*

(CC1)* $\sum_{i=1}^{r} A_i = J$.

(CC2) There is a subset of $\mathcal{A}$ with sum I.*

(CC3) $\mathcal{A}$ is closed under taking transposes.*

(CC4) For $i, j, k \in \{1, \ldots, r\}$, there are numbers p_{ij}^{k} such that*

$$A_i A_j = \sum_{k=1}^{r} p_{ij}^{k} A_k.$$

Let $V(\mathcal{A})$ be the span of the set $\mathcal{A}$ over the field $\mathbb{C}$. Then condition (CC4*) asserts that $V(\mathcal{A})$ is multiplicatively closed, that is, it is a $\mathbb{C}$-algebra. We call it the *basis algebra* of the coherent configuration. Condition (CC2*) says that this algebra contains the identity, and (CC3*) that it is closed under transposition.

Now the following conditions on a permutation g of Ω are equivalent. (Here $P(g)$ is the permutation matrix representing g.)

- g is an automorphism of the coherent configuration $(\Omega, \mathcal{R})$;
- g preserves every relation $R \in \mathcal{R}$;
- $P(g)$ commutes with every basis matrix $A(R)$ for $R \in \mathcal{R}$;
- $P(g)$ commutes with $V(\mathcal{A})$.

Conversely, if the coherent configuration is associated with a group G, then its basis algebra consists of all complex matrices which commute with every permutation matrix $P(g)$ for $g \in G$. For this reason, an alternative name for the basis algebra (in the group case) is the *centraliser algebra* of G.

3.2 Algebraic theory

What can we say about the structure of the basis algebra?

Theorem 3.2 *The basis algebra of a coherent configuration is isomorphic to a direct sum of complete matrix algebras over $\mathbb{C}$. If the coherent configuration is associated with a permutation group G, then the degrees of these complete matrix algebras are equal to the multiplicities of the irreducible constituents of the permutation character of G.*

Proof. We first show that $V(\mathcal{A})$ is semisimple (that is, a direct sum of minimal ideals). For this, we observe that $V(\mathcal{A})$ is closed under taking the conjugate transpose (since this is true for the basis $\mathcal{A}$). Now consider the positive definite inner product $[A, B]$ on the space of $n \times n$ complex matrices, given by

$$[A, B] = \mathrm{Trace}(A\overline{B}^{\top}) = \sum_{i=1}^{n}\sum_{j=1}^{n} a_{ij}\overline{b_{ij}}.$$

The restriction of this inner product to $V(\mathcal{A})$ is still positive definite. It will suffice to show that, for any ideal J, the space $J^{\perp}$ is also an ideal; for then we obtain the required decomposition by induction. So let J be an ideal, and take $B \in J^{\perp}$. We have to show that $XB, BX \in J^{\perp}$ for any $X \in V(\mathcal{A})$.

Now

$$\begin{aligned}
[A, XB] &= \mathrm{Trace}(A\overline{(XB)}^{\top}) \\
&= \mathrm{Trace}(A\overline{B}^{\top}\overline{X}^{\top}) \\
&= \mathrm{Trace}(\overline{X}^{\top}A\overline{B}^{\top}) \\
&= [\overline{X}^{\top}A, B] \\
&= 0,
\end{aligned}$$

because $\overline{X}^{\top} \in V(\mathcal{A})$ and J is an ideal. Hence $XB \in J^{\perp}$. The proof that $BX \in J^{\perp}$ is similar but easier.

Now a semisimple algebra over $\mathbb{C}$ is a direct sum of complete matrix algebras over $\mathbb{C}$, by Wedderburn's Theorem (see, for example, Lang [111], p. 445).

In the case of coherent configurations associated with groups, we use the argument of the last chapter. Let P be the permutation representation, and let it decompose as the sum of e_i copies of the irreducible representation M_i of degree f_i, for $i = 1, \ldots, k$. Then P is equivalent to the representation

$$\begin{bmatrix}
I_{e_1} \otimes M_1 & O & \ldots & O \\
O & I_{e_2} \otimes M_2 & \ldots & O \\
\vdots & \vdots & \ddots & \vdots \\
O & O & \ldots & I_{e_k} \otimes M_k
\end{bmatrix}.$$

The centraliser algebra then consists of all matrices with shape

$$\begin{bmatrix} A_1 \otimes I_{f_1} & O & \dots & O \\ O & A_2 \otimes I_{f_2} & \dots & O \\ \vdots & \vdots & \ddots & \vdots \\ O & O & \dots & A_k \otimes I_{f_k} \end{bmatrix},$$

where A_i is an arbitrary $e_i \times e_i$ matrix.

Corollary 3.3 *The basis algebra of a coherent configuration associated with a permutation group G is commutative if and only if the permutation character of G is* multiplicity-free *(that is, its irreducible constituents occur with multiplicity 1).*

There is another algebra associated with a coherent configuration, as follows. For $1 \le j \le r$, let P_i be the $r \times r$ matrix with (s,t) entry $(P_j)_{st} = p^t_{sj}$. The matrices P_j are the *intersection matrices* of the configuration (as their entries are the intersection numbers).

Theorem 3.4 *(a) The span of the set $\{P_i : 1 \le i \le r\}$ of matrices over $\mathbb{C}$ is an algebra, called the* intersection algebra *of the coherent configuration.*

(b) The map $A_i \mapsto P_i$, extended linearly, is a $\mathbb{C}$-algebra isomorphism from the basis algebra to the intersection algebra.

Proof. First, we prove the identity

$$\sum_{s=1}^{r} p^s_{ij} p^u_{sk} = \sum_{t=1}^{r} p^u_{it} p^t_{jk}.$$

This can be shown in two different ways. First, given $(\alpha,\beta) \in R_u$, it counts the choices of two points γ,δ such that $(\alpha,\gamma) \in R_i$, $(\gamma,\delta) \in R_j$, and $(\delta,\beta) \in R_k$. For there are p^u_{sk} points δ with $(\alpha,\delta) \in R_s$, and for each such point δ, there are p^s_{ij} choices of γ; multiplying and summing gives the result. The right-hand side is obtained similarly, first choosing γ with $(\gamma,\beta) \in R_t$.

Alternatively, this identity follows from the associative law: the two sides are the coefficient of A_u in the products $(A_iA_j)A_k$ and $A_i(A_jA_k)$ respectively.

Now examine this equation. It asserts that

$$(P_jP_k)_{iu} = \sum_{s=1}^{r} (P_j)_{is}(P_k)_{su} = \sum_{t=1}^{r} p^t_{jk}(P_t)_{iu},$$

or

$$P_jP_k = \sum_{t=1}^{r} p^t_{jk} P_t.$$

Figure 3.1: The associative law

This shows that the span of the intersection matrices is an algebra. Also, the relations defining the basis algebra are

$$A_jA_k = \sum_{t=1}^{r} p_{jk}^{t} A_t,$$

and so the map $A_j \mapsto P_j$ is an algebra homomorphism.

To show that it is an isomorphism, it suffices to prove that the intersection matrices P_j are linearly independent. To prove this, we look at the first m rows. For each orbital R_j, there are two orbits Ω_a and Ω_b such that the ordered pairs in R_j have first component in Ω_a and second component in Ω_b. Then the only non-zero element in the first m rows of P_j occurs in row a and column j. The linear independence is now clear.

This result was proved (in the commutative case) by Bose and Mesner [23]. Higman attributes the general result to Wielandt.

The main advantage of using the intersection algebra rather than the basis algebra is that the matrices are usually much smaller: their size is equal to the rank rather than the number of points. Note that two non-isomorphic coherent configurations may have the same intersection matrices. See Exercise 3.9 for an example.

The numbers e_i and f_i can in principle be determined from the intersection numbers p_{ij}^{k}. (The e_i are the degrees of irreducible matrix representations of the intersection algebra, and the f_i are their multiplicities in the basis algebra.) However, the details of this process are clearer in the case where the algebras are commutative, which we treat next.

3.3 Association schemes

Although the concept of a coherent configuration is tailor-made for describing the orbits of a permutation group on pairs of points, the roots of the subject lie in a different area, namely statistics (and in particular, experimental design). R. C. Bose and his co-workers developed the theory of association schemes in connection with *partially balanced designs* (see [24]). To Bose's school, an association scheme is a coherent configuration $(\Omega, \mathcal{R})$ in which *all the relations in* $\mathcal{R}$ *are symmetric.* This is a very natural assumption in statistics, for two reasons. First, the relations on the set of treatments are often defined by *concurrences*, the concurrence of a pair of treatments being the number of blocks of the design in which they both occur. Second, if all the basis matrices are symmetric, then they commute (see Exercise 3.2), and so can be simultaneously diagonalised. Thus the real vector space $\mathbb{R}\Omega$ can be decomposed into orthogonal subspaces which consist of eigenvectors for all the basis matrices. Statisticians routinely have to decompose real vector spaces – indeed, the data from an experiment is a real vector, and its components measure treatment and interaction effects – and so this situation is convenient.

Later, Delsarte [59] showed that association schemes form a unifying principle in the theories of codes and designs. He proved many powerful results from this point of view. He observed that his proofs do not require that the relations are symmetric, but only that the basis matrices commute. (However, all important examples in his work are symmetric.)

We will say that a coherent configuration is a *commutative association scheme* if the basis matrices commute; it is a *symmetric association scheme* if all the basis matrices are symmetric. As we observed above, a symmetric association scheme is commutative. For the group case, we recognise association schemes as follows.

Theorem 3.5 *Let G be a permutation group on Ω, having permutation character π.*

(a) The coherent configuration of G is a commutative association scheme if and only if π is multiplicity-free.

(b) The coherent configuration of G is a symmetric association scheme if and only if π is multiplicity-free and all its irreducible constituents have Frobenius–Schur index $+1$.

Note that, in an association scheme, condition (CC2) is strengthened to:

(CC2†) The diagonal of Ω is a single relation in $\mathcal{R}$.

For this reason it is often convenient to change the numbering of the relations, using $0, \dots, r-1$ instead of $1, \dots, r$, with the diagonal relation being R_0.

Two important examples of association schemes (especially in Delsarte's theory) are the following.

- The *Hamming scheme* $H(n, q)$: the points are all n-tuples of elements from an alphabet Q of size q. For $0 \leq i \leq n$, two n-tuples satisfy the ith relation R_i if and only if their *Hamming distance* is i: that is, they differ in i places and agree in the remaining $n-i$ places. The Hamming scheme is the coherent configuration of the group $S_q \operatorname{Wr} S_n$ in its *product action* (which will be examined in more detail in the next chapter).

- The *Johnson scheme* $J(v, k)$: the points are all the k-element subsets of a set of size v. We assume that $k \leq v/2$. For $0 \leq i \leq k$, two subsets satisfy the ith relation R_i if their intersection has cardinality $k - i$. The Johnson scheme is the coherent configuration of the group S_v in its action on the set of k-subsets of $\{1, \ldots, v\}$.

Another type consists of the conjugacy class schemes of groups. These are the schemes which arise from the diagonal action

$$(g, h) : x \mapsto g^{-1}xh$$

of $G \times G$ on G, for any group G. The name arises because the classes $R_i(1)$ are the conjugacy classes of G.

There is a particularly important class of symmetric association schemes, derived from graphs. Let Γ be a connected graph on the vertex set Ω, having diameter d. Let $d(x, y)$ be the distance between the vertices x and y of Γ. We let $\Gamma_i(x)$ denote the set of all vertices whose distance from the vertex x is i, for $0 \leq i \leq d$ (so that $\Gamma_0(x) = \{x\}$ and $\Gamma_1(x)$ is the set of neighbours of x).

- We say that Γ is *distance-transitive* if, for $0 \leq i \leq d$, the group $G = \operatorname{Aut}(\Gamma)$ acts transitively on the set of ordered pairs of vertices at distance i.

- We say that Γ is *distance-regular* if there are numbers c_i, a_i, b_i for $0 \leq i \leq d$ such that, if $d(x, y) = i$, then the number of neighbours of y which lie at distance $i-1, i, i+1$ respectively from x are c_i, a_i, b_i respectively.

(Note that, for a distance-regular graph, the parameters c_0 and b_d are undefined.)

Theorem 3.6 *(a) A distance-transitive graph is distance-regular.*

(b) If Γ is distance-regular with diameter d, then the relations

$$R_i = \{(x, y) : d(x, y) = i\}$$

for $0 \leq i \leq d$ form a symmetric association scheme on the vertex set of Γ.

Proof. (a) Let (x, y) and (x', y') be two pairs of vertices at distance i. If $g \in \mathrm{Aut}(\Gamma)$ maps (x, y) to (x', y'), then g maps the set of neighbours of y at distance j from x to the set of neighbours of y' at distance j from x', for $j = i - 1, i, i + 1$; so the numbers of vertices in these sets are the same.

(b) Let A_i be the basis matrix of the relation R_i. Using the definition of distance-regularity, we see that

$$A_i A_1 = b_{i-1} A_{i-1} + a_i A_i + c_{i+1} A_{i+1},$$

with the convention that $A_{-1} = A_{d+1} = O$. Now the connectedness of Γ shows that $c_i > 0$ for $i = 0, \ldots, d - 1$. Then an induction based on the displayed equation shows that, for all i, we can express A_1^i as a linear combination of the basis matrices; and, conversely, each basis matrix A_i is a polynomial of degree i in A_1. Moreover, A_1 satisfies a polynomial of degree $d + 1$, obtained from the equation

$$A_1 A_d = b_{d-1} A_{d-1} + a_d A_d;$$

using this, we can express any product $A_i A_j$ as a polynomial of degree at most d in A_1. It follows that the span of the set of basis matrices is closed under multiplication, and we do indeed have a symmetric association scheme.

Remark. Both the Hamming and the Johnson schemes arise from distance-transitive graphs in this way. The numbering in our definition of these schemes was chosen so that A_1 is the adjacency matrix of the graph, and the relation R_i corresponds to distance i in the graph.

The valency n_i associated with the relation R_i in a distance-regular graph is the (constant) number of vertices at distance i from a given vertex. These numbers can be calculated as follows. Counting in two ways the edges between $\Gamma_i(x)$ and $\Gamma_{i+1}(x)$, we see that

$$n_i b_i = n_{i+1} c_{i+1}.$$

Hence by induction (since $n_0 = 1$), we have

$$n_i = b_0 b_1 \cdots b_{i-1} / c_1 c_2 \cdots c_i.$$

In particular, $n_1 = b_0$ is the valency of the graph Γ, usually denoted by k; and $a_0 = 0, c_1 = 1$. We note in passing that $c_i + a_i + b_i = k$ for $i = 0, \ldots, d$ (taking the undefined c_0 and b_d to be zero).

The parameters of distance-regular graphs satisfy a number of restrictions. We mention just one, which will be used later in the proof of Smith's Theorem 3.19.

Theorem 3.7 *The parameters of a distance regular graph satisfy*

(a) $c_1 \leq c_2 \leq \ldots \leq c_d$;

(b) $b_0 \geq b_1 \geq \ldots \geq b_{d-1}$.

Proof. (a) Suppose that $d(x,y) = i$, with $2 \leq i \leq d$. Choose a point u adjacent to x on a shortest path from x to y (so that $d(x,u) = 1$ and $d(y,u) = i-1$). Then any vertex z adjacent to y at distance $i-2$ from u lies at distance $i-1$ from x. (Its distance is not greater than $d(x,u)+d(u,z) = i-1$; and not smaller than $d(x,y) - d(y,z) = i-1$.) So the set of vertices counted by c_{i-1} (relative to u and y) is a subset of the set counted by c_i (relative to x and y), and $c_{i-1} \leq c_i$.

(b) The proof is similar.

From this result and the formula

$$n_i = b_0 b_1 \cdots b_{i-1} / c_1 c_2 \cdots c_i$$

for the number of vertices at distance i from x, we see that the numbers n_i are *unimodal*: they increase, then remain constant, then decrease (where any of these steps may be absent).

For more information on distance-regular graphs, see the book by Brouwer, Cohen and Neumaier [25]. On association schemes in general, see Bailey [12] (this book gives several different practical techniques for finding the eigenvalues and multiplicities of basis matrices), or Bannai and Ito [14]. A valuable Internet resource is the database maintained by Andries Brouwer: his instructions for using it follow.

```
In order to obtain the spectrum of [and possibly other
information about] strongly regular or distance-regular
graphs, send email to
        aeb@cwi.nl
with a subject line (in the header)
        Subject: exec drg
followed by a body consisting of zero or more lines
of the form
        drg d=D B[0],B[1],...,B[d-1]:C[1],...,C[d]
where D is the diameter, and the B[i] and C[i] are
the usual intersection numbers.
```

3.4 Algebra of association schemes

The basis algebra of an association scheme was first studied by Bose and Mesner [23]; for this reason, it is often referred to as the *Bose–Mesner algebra* of the scheme.

If a complete matrix algebra over $\mathbb{C}$ is commutative, then it has degree 1. So Theorem 3.2 shows that the basis matrices of an association scheme can be simultaneously diagonalised. This can be seen another way. The set $\mathcal{A}$ of basis matrices consists of commuting symmetric real matrices (in the symmetric case) or commuting normal complex matrices (in general), and so can be simultaneously diagonalised by a real orthogonal matrix (in the symmetric case) or by a complex unitary matrix (in general). (A complex matrix is *normal* if it commutes with its conjugate transpose.) So the eigenspaces of the basis matrices are orthogonal (with respect to the standard real or complex inner product).

The first such eigenspace is easy to describe. Each basis matrix A_i has constant row and column sums $n_i = p^0_{ii^*}$. So the all-1 vector $\mathbf{1}$ is a common eigenvector, satisfying $A_i\mathbf{1} = n_i\mathbf{1}$.

Number the common eigenspaces as $W_0, W_1, \ldots, W_{r-1}$, in any manner provided that $W_0 = \langle \mathbf{1} \rangle$. As noted above, these subspaces are mutually orthogonal. The basis algebra $V(\mathcal{A})$ consists of all matrices which preserve these subspaces and act as a scalar on each one. In particular, the *orthogonal projection* E_j onto W_j is an element of the adjacency algebra, for each j. If a matrix $A \in V(\mathcal{A})$ has eigenvalue $\pi_j(A)$ on the space W_j, then we have the *spectral decomposition*

$$A = \sum_{j=0}^{r-1} \pi_j(A) E_j.$$

Note that $\pi_j(A)$ is also an eigenvalue of the image of A in the intersection algebra.

We use the notation $\pi_j(A_i) = p_i(j)$ for $0 \leq i, j \leq r-1$. We now show how to compute the dimensions of the spaces W_j (which, in the group case, are the degrees of the irreducible constituents of the permutation character). We also use the notation i^* for the index of the relation paired with i, so that $A_{i^*} = A_i^{\top}$. Note that $p_{i^*}(j) = \overline{p_i(j)}$.

Theorem 3.8 *Let u_j be a common left eigenvector of the intersection algebra of an association scheme, and v_j a common right eigenvector with the same eigenvalues. Normalise these vectors so that $(u_j)_0 = (v_j)_0 = 1$. Then the dimension of the corresponding eigenspace of the basis algebra is given by*

$$f_j = \frac{|\Omega|}{u_j v_j}.$$

(The denominator is the product of the row vector u_j and the column vector v_j, that is, $\sum_{i=0}^{r-1}(u_j)_i(v_j)_i$.)

A proof can be found in [43], p. 202, for symmetric association schemes; the extra generality gives no difficulty. The proof shows that

$$\begin{aligned} u_j &= (p_{0\bullet}(j)/n_0, p_{1\bullet}(j)/n_1, \ldots, p_{r-1\bullet}(j)/n_{r-1}), \\ v_j &= (p_0(j), p_1(j), \ldots, p_{r-1}(j))^{\top}, \end{aligned}$$

so we have the formula

$$f_j = \frac{n}{\sum_{i=0}^{r-1} |p_i(j)|^2/n_i}.$$

Delsarte defined the *P and Q matrices* (or *first and second eigenmatrices*) of an association scheme. The matrices defined here are the transposes of Delsarte's; this is a little more convenient. We set

$$(P)_{ij} = p_i(j), \qquad (Q)_{ij} = f_i p_{j\bullet}(i)/n_j.$$

The proof of Theorem 3.8 shows that:

Theorem 3.9 $PQ = QP = nI$.

The Q matrix is a generalisation of the character table of a group (see Exercise 3.10), and Theorem 3.9 generalises the orthogonality relations for the character table which we met in the last chapter. Note that P and $n^{-1}Q$ are the transition matrices between the two bases $\{A_0, \ldots, A_{r-1}\}$ and $\{E_0, \ldots, E_{r-1}\}$ of $V(\mathcal{A})$: specifically,

$$A_i = \sum_{j=0}^{r-1} (P)_{ij} E_j, \quad E_i = n^{-1} \sum_{j=1}^{r-1} (Q)_{ij} A_j.$$

Of course, the multiplicities f_j are positive integers. This gives a very powerful necessary condition on the parameters for an association scheme to exist, the so-called *integrality condition*: these r algebraic functions of the parameters, calculated from the intersection matrices, must be positive integers.

For an early application of this technique, which was of great importance for group theory, see [73].

From this, we can derive a simple necessary condition for the existence of an association scheme, *Frame's condition*:

Theorem 3.10 *If an association scheme with n points and s classes (that is, rank $s+1$) has valencies $n_1,\ldots,n_s$ and multiplicities $f_1,\ldots,f_s$, then*

$$F = n^{s-1}n_1\cdots n_s/f_1\cdots f_s$$

is an integer. Moreover, if all the eigenvalues of the scheme are rational, then F is a square. In particular this holds if all the f_i are different.

Proof. First we note that $\det(P)$ is a multiple of n: for the sum of the rows is $(n,0,\ldots,0)$, since these are the eigenvalues of the all-1 matrix J. More precisely, $\det(P)/n$ is an algebraic integer, which is a rational integer if the entries of P are rational integers.

Now we have $Q = \operatorname{diag}(f_i)\overline{P}^{\top}\operatorname{diag}(n_i^{-1})$, and $PQ = nI$. So

$$\left(\prod_{i=1}^{s} f_i\right)|\det(P)|^2 \Big/ \left(\prod_{i=1}^{s} n_i\right) = n^{s+1}.$$

Thus $F = |\det(P)/n|^2$, and the theorem is proved, except for the last sentence. But, if there is an irrational eigenvalue, then its algebraic conjugates are also eigenvalues, and have the same multiplicity.

For distance-regular graphs, we can do the calculation another way. Define a sequence of polynomials by

$$\begin{aligned} &f_0(x) = 1, \qquad f_1(x) = x,\\ &xf_i(x) = b_{i-1}f_{i-1}(x) + a_i f_i(x) + c_{i+1}f_{i+1}(x) \qquad (1 \le i \le d-1). \end{aligned}$$

Our earlier analysis shows that $A_i = f_i(A_1)$ for $0 \le i \le d$. Also, since $A_1A_d = b_{d-1}A_{d-1} + a_dA_d$, we see that A_1 satisfies the polynomial

$$(x-a_d)f_d(x) - b_{d-1}f_{d-1}(x) = 0.$$

Since A_1 generates the basis algebra, this is its minimal polynomial. It is also the minimal polynomial of the intersection matrix P_1, which is *tridiagonal*:

$$P_1 = \begin{pmatrix} 0 & 1 & 0 & \cdots & \cdots & \cdots\\ k & a_1 & c_2 & 0 & \cdots & \cdots\\ 0 & b_1 & a_2 & c_3 & 0 & \cdots\\ \ddots & \ddots & \ddots & \ddots & \ddots & \ddots\\ \cdots & \cdots & 0 & b_{d-2} & a_{d-1} & c_d\\ \cdots & \cdots & \cdots & 0 & b_{d-1} & a_d \end{pmatrix}.$$

The eigenvalues of A_1 are the roots of this minimal polynomial $m(x)$. This gives another way to calculate their multiplicities.

Let θ be a root of $m(x) = 0$, and let $f(x) = m(x)/(x-\theta)$. Then $g(x) = f(x)/f(\theta)$ is a polynomial with the property that $f(\theta) = 1$ while $f(\phi) = 0$ for any other root ϕ of $m(x)$. Hence $g(A) = E$ is the idempotent representing orthogonal projection onto the θ eigenspace of A_1, and the multiplicity of θ is equal to the trace of $g(A_1)$. Now Trace(A_1^k) is equal to the number of closed walks of length k in the graph, which can be calculated from the parameters. (For example, Trace$(A_1) = 0$, Trace$(A_1^2) = nk$, and Trace$(A_1^3) = nka_1$.) Note also that $f(\theta) = m'(\theta)$ (since differentiating $m(x)$ gives, by Leibniz' Rule, $d+1$ terms, all but one of which has θ as a root); this simplifies the calculation.

Many of the most powerful results about non-existence of distance-regular graphs arise from the fact that good estimates can be made for the eigenvalues of tridiagonal matrices, and hence for their multiplicities.

3.5 Example: Strongly regular graphs

A *strongly regular graph* is a graph Γ on the vertex set Ω having the property that $(\Omega, \{R_0, R_1, R_2\})$ is an association scheme, where R_0, R_1, R_2 are the relations of equality, adjacency, and non-adjacency respectively.

Theorem 3.11 *(a) A disconnected strongly regular graph is a disjoint union of complete graphs of the same size, and conversely.*

(b) A connected strongly regular graph is a distance-regular graph of diameter 2, and conversely.

We will not consider disconnected strongly regular graphs further. Let Γ be a connected strongly regular graph. By convention, we use the letters k, λ, μ for the parameters b_0, a_1, c_2. Thus, the graph has valency k, and the number of common neighbours of two vertices x, y is λ (if x and y are adjacent) or μ (if not). Also, we let l be the number of vertices not adjacent to a given vertex. Thus, the total number of vertices is $n = 1 + k + l$. Double-counting edges between the neighbours and the non-neighbours of x, we find that $k(k-\lambda-1) = l\mu$. (This is just the formula $k_2 = b_0b_1/c_2$ for a distance-regular graph.) The intersection matrices of Γ are

$$P_0 = \begin{pmatrix} 1 & 0 & 0 \\ 0 & 1 & 0 \\ 0 & 0 & 1 \end{pmatrix}, \quad P_1 = \begin{pmatrix} 0 & 1 & 0 \\ k & \lambda & \mu \\ 0 & k-\lambda-1 & k-\mu \end{pmatrix},$$

$$P_2 = \begin{pmatrix} 0 & 0 & 1 \\ 0 & k-\lambda-1 & k-\mu \\ l & l-k+\lambda+1 & l-k+\mu-1 \end{pmatrix}.$$

The minimal polynomial of P_1 is $(x-k)(x^2-(\lambda-\mu)x+(k-\mu))$. We see that P_1 has an eigenvalue k (the column sums); the left and right eigenvectors are

$(1,1,1)$ and $(1,k,l)$, and so the multiplicity is $n/(1+k+l)=1$, as expected. The other eigenvalues r,s of P_1 satisfy $x^2-(\lambda-\mu)x+(k-\mu)=0$, or

$$r,s=\frac{(\lambda-\mu)\pm\sqrt{(\lambda-\mu)^2+4(k-\mu)}}{2}.$$

The multiplicities f,g can be computed most easily as follows. We have

$$f+g=k+l$$

and

$$k+fr+gs=\operatorname{Trace}(A_1)=0.$$

Solving these equations for f and g gives

$$f,g=\frac{1}{2}\left((k+l)\pm\frac{(k+l)(\mu-\lambda)-2k}{\sqrt{(\lambda-\mu)^2+4(k-\mu)}}\right).$$

These numbers are positive integers. We distinguish two cases:

Case I: $(k+l)(\mu-\lambda)-2k=0$. Since $k+l>k$ and $k+l$ divides $2k$, we have $l=k$ and $\mu=\lambda+1$. Now the equation $k(k-\lambda-1)=l\mu$ gives us $k-\mu=\mu$, or $k=2\mu$. We have $f=g=k=l$.

Case II: $(k+l)(\mu-\lambda)-2k\neq 0$. Now the expression under the square root sign must be a perfect square, say u^2; and u divides the numerator in the expressions for f and g, the quotient being congruent to $k+l$ modulo 2. In this case, the eigenvalues r,s are rational, and hence are integers, with $r>0$ and $s<0$. It is often convenient to express the other parameters in terms of k,r,s:

$$\lambda=k+r+s+rs,\quad \mu=k+rs,\quad l=\frac{-k(r+1)(s+1)}{k+rs}.$$

The expressions for l,f,g give further divisibility conditions the parameters must satisfy.

We give an example. A *Moore graph of diameter* 2 is a regular connected graph of diameter 2 containing no cycles of length 3 or 4. Thus, it is a strongly regular graph with $\lambda=0$ and $\mu=1$. *Which Moore graphs exist?* The following analysis is due to Hoffman and Singleton [94].

If Case I occurs, then $k=l=2\mu=2$, so $n=5$, and the graph is a pentagon (which is indeed a Moore graph). Suppose that we are in Case II. We have

$$r+s=\lambda-\mu=-1,$$

so $s = -1 - r$. Then $k = 1 - rs = r^2 + r + 1$, $l = r(r+1)(r^2+r+1)$, and

$$f, g = \frac{1}{2}\left((r^2+r+1)^2 \pm \frac{(r^2+r+1)(r^2+r-1)}{2r+1}\right).$$

Thus, $2r+1$ divides $(r^2+r+1)(r^2+r-1)$. So $2r+1$ divides $3 \cdot 5 = 15$. (Multiply $(r^2+r+1)(r^2+r-1)$ by 16 and put $2r \equiv -1$.) The possibilities are:

r	0	1	2	7
k	1	3	7	57
n	2	10	50	3250

The first case does not give a legal Moore graph (it is a single edge, and is indeed regular with no cycles of length 3 or 4, but does not have diameter 2.) The graphs in the second and third cases exist and are unique. The second is the *Petersen graph*, shown in Figure 3.2. The third is the *Hoffman–Singleton graph*: see the next section for a construction (see also Exercise 3.13). It is not known whether a Moore graph of valency 57 exists: we show in the following section that no such graph can have a transitive automorphism group.

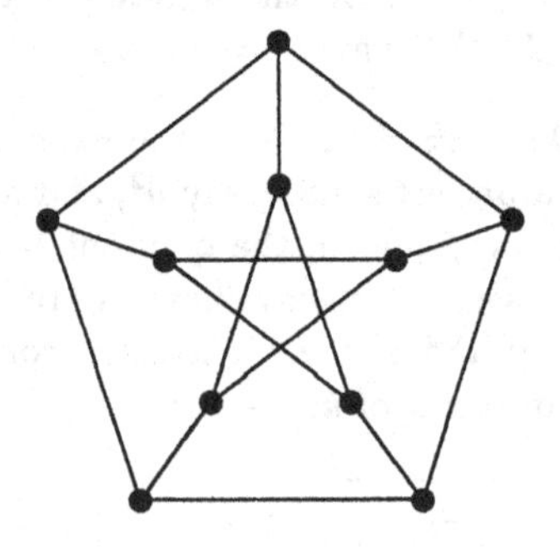

Figure 3.2: The Petersen graph

Theorem 3.12 *A Moore graph of diameter* 2 *must have* 5, 10, 50, *or possibly* 3250 *vertices, and valency* 2, 3, 7, *or possibly* 57 *respectively. Graphs in the first three cases exist and are unique up to isomorphism, and their automorphism groups are transitive and have rank* 3 *on vertices.*

The automorphism groups are D_{10}, S_5, and $\mathrm{P\Sigma U}(3, 5^2)$ respectively.

3.6 The Hoffman–Singleton graph

The simplest construction of the Hoffman–Singleton graph follows. Recall from Exercise 1.35 that there are 30 ways to construct a projective plane on the set $\{1,\ldots,7\}$, falling into two orbits of 15 under the alternating group A_7. Moreover, any two 3-sets which intersect in one point are lines of two of these planes, one in each orbit (since they differ by a transposition), and any 3-set is a line in three planes in each orbit.

The vertices of the graph are of two types: the 3-subsets of $\{1,\ldots,7\}$, and an A_7-orbit of projective planes. So there are $\binom{7}{3}+15=50$ vertices. Two 3-sets are joined if they are disjoint; a 3-set is joined to a projective plane if it is a line of that plane; and two projective planes are never joined. We show that the graph has valency 7 and diameter 2; it follows that it is a Moore graph.

Any 3-set is disjoint from four 3-sets and (as already noted) is a line in three projective planes. Any projective plane has seven lines. So the valency is 7.

Take two non-adjacent vertices x, y: we must show that they have a common neighbour. There are several cases:

- x and y are 3-sets. If they meet in two points, there is a 3-set disjoint from both; if they meet in one point, they are both lines in a projective plane.
- x is a 3-set and y a projective plane. Then x is a triangle in y, so there is a line of y disjoint from x.
- x and y are projective planes. The fact that they have a common line follows from the 2-transitivity of A_7 on its orbit (since *some* pair of planes share a line).

In this section, we use GAP to construct this graph and investigate its automorphism group. We also construct two other distance-transitive graphs. We will use the GAP 'share package' GRAPE [169], which is useful for computing with graphs and groups. It is loaded using the command

```
gap> RequirePackage("grape");
```

A couple of notes about GAP first. The commands that follow are fairly lengthy, and you may prefer to write a file containing them, and then get GAP to read the file. If you do this, output is suppressed, and I have adopted the same convention here.

Also, we must understand the distinction between a set and a list. Lists are basic data structures in GAP, but a list can contain gaps or repetitions, and the elements can be in any order. A set has its elements arranged in order without gaps or repetitions. The function `Set` converts a list into a set. When we apply a permutation to a set, the result may be a list and need to be converted back to a set. The code `List(L, x -> f(x))` replaces the elements of the list `L` with their images under the function `f`.

First we construct the Hoffman–Singleton graph. Since projective planes are sets of sets, on which we can't act directly with GAP, we first build a recursive action of permutations on sets nested to any depth.

```
OnSetsRecursive:=function(x,g)
  if not IsSet(x) then
    return x^g;                        # this is the action of
                                       # permutation g on point x
  else
    return Set(List(x, y -> OnSetsRecursive(y,g)));
  fi;
end;
```

Now we define adjacency in the Hoffman–Singleton graph as a boolean function, according to the recipe given earlier.

```
HSAdjacent:=function(x,y)
  if IsInt(x[1]) then                  # x is a 3-set
    if IsInt(y[1]) then                # y is a 3-set
      return Intersection(x,y)=[]; # join if disjoint
    else                               # or y is a projective plane
      return x in y;                   # join if x is a line of y
    fi;
  else                                 # or x is a projective plane
    if IsInt(y[1]) then                # and y is a 3-set
      return y in x;                   # join if y is a line of x
    else                               # or y is a projective plane
      return false;                    # don't join
    fi;
  fi;
end;
```

For convenience we write down a projective plane.

```
Pi:=Set([[1,2,4],[2,3,5],[3,4,6],[4,5,7],
     [1,5,6],[2,6,7],[1,3,7]]);
```

Now we invoke the general graph constructor from `GRAPE`. This takes four arguments: a group G; a list of elements in a set acted on by the group; the function giving the action; and a G-invariant function defining adjacency. It produces the graph with this adjacency rule on the union of orbits of the elements in the list.

```
HofSingGraph:=Graph(AlternatingGroup(7),
                    [[1,2,3], Pi],
                    OnSetsRecursive,
                    HSAdjacent);

HofSingGroup:=AutGroupGraph(HofSingGraph);
```

Now properties of the graph and its group, such as the facts that the graph is a Moore graph, and that the group has rank 3 on the vertices, can be checked at leisure. There are built-in functions in GRAPE to check distance-regularity (`IsDistanceRegular`) and to find the parameters of a distance-regular graph (`GlobalParameters`). By displaying the composition factors of the automorphism group, we find that it has a subgroup of index 2 which is the simple group $\mathrm{PSU}(3,5^2)$ (or `U(3,5)` in GAP notation).

The philosophy of GRAPE is that a graph comes equipped with a group of automorphisms which is used to speed up calculations. As we have defined the Hoffman–Singleton graph, it is equipped with the group A_7. To do any computation requiring serious searching in this graph, it is wise to 'update' the group to the full automorphism group, which (being 100 times larger) will cause the calculations to run much faster. This is done by the command

```
HofSingGraph := NewGroupGraph(HofSingGroup,HofSingGraph);
```

We go on to construct two more distance-transitive graphs. Each graph has 100 vertices, these being the images of a 15-coclique (which can be taken as the set of projective planes in the original description). The next function builds this coclique as a set of vertices of the graph. Here we invoke another useful GRAPE feature. While the vertex set of the graph we've constructed is just `[1..50]`, the origins of the vertices (as 3-sets or projective planes) are remembered, and can be recovered by the command `VertexName`. We filter out the vertices whose names are not sets of integers.

```
IsPlane:=function(x)
  return not IsInt(VertexName(HofSingGraph,x)[1]);
end;

Coclique:=Set(Filtered(Vertices(HofSingGraph),IsPlane));
```

Now it can be checked that there are 100 images of this coclique, and that any two meet in 0, 3, 5 or 8 points. The numbers meeting a given coclique in

0, 3, 5, 8 points are 7, 35, 42, 15 respectively. This can be established with the following code, which intersects each image of `Coclique` with `Coclique`, and collects the sizes with their numbers of occurrences.

```
Cocliques:=Orbit(HofSingGroup, Coclique, OnSets);

IntList:=Collected(List(Cocliques,
         x -> Size(Intersection(x,Coclique))));
```

It turns out that the graph obtained by joining cocliques with intersection 0 is the disjoint union of two copies of the Hoffman–Singleton graph; the graph for intersection 8 is bipartite distance-transitive with diameter 4; and the graph for intersection 0 or 8 is the *Higman–Sims graph*, a strongly regular graph with rank 3 automorphism group. All these facts are easily checked when the graphs are constructed. The constructions all work the same way using the graph constructor with different joining rules.

```
Joined:=function(x,y,L)    # L is a list of intersection sizes
  return Size(Intersection(x,y)) in L;
end;

J0:=function(x,y)
  return Joined(x,y,[0]);
end;

Graph0:=Graph(HofSingGroup, [Coclique], OnSets, J0);

J8:=function(x,y)
  return Joined(x,y,[8]);
end;

Graph8:=Graph(HofSingGroup, [Coclique], OnSets, J8);

J08:=function(x,y)
  return Joined(x,y,[0,8]);
end;

Graph08:=Graph(HofSingGroup, [Coclique], OnSets, J08);
```

The claim made earlier about `Graph0` can be verified by using the functions `ConnectedComponent` (which returns the *vertex set* of the connected component of a graph containing a given vertex), `InducedSubgraph` (which constructs the induced subgraph of a graph on a given set of vertices) and `IsIsomorphicGraph` (which tests the isomorphism of two graphs).

The automorphism group of the graph `Graph08` has a subgroup of index 2 which is the sporadic Higman–Sims simple group. If you wish to continue the exercise, you can show that the Higman–Sims group contains the automorphism group of `Graph8` (this group being isomorphic to $\mathrm{P\Sigma L}(3,5^2)$, the automorphism group of the Hoffman–Singleton graph), and that the action of the Higman–Sims group on the 176 cosets of this subgroup is 2-transitive.

Another construction of the Higman–Sims graph using GAP can be found at

`http://www-gap.dcs.st-and.ac.uk/~gap/Talks/bcc.html`

Remark. The output of `GRAPE`'s `GlobalParameters` function has to be rearranged for Brouwer's DRG finder (see Section 3.3). For example, we have

```
gap> GlobalParameters(Graph8);
[ [ 0, 0, 15 ], [ 1, 0, 14 ], [ 5, 0, 10 ],
  [ 12, 0, 3 ], [ 15, 0, 0 ] ]
```

and on the basis of this we could send a message

```
drg d=4  15, 14, 10, 3 : 1, 5, 12, 15
```

to Brouwer's daemon for further information.

We have only explored a small subset of the GAP resources for dealing with groups, graphs, etc. See the documentation to find out what is available.

3.7 Automorphisms

The algebraic machinery of association schemes gives us more than just a non-existence test based on the integrality of character degrees. We can compute character values, and so prove non-existence of specific automorphisms, as we now show.

Consider an association scheme with basis matrices $A_0, \ldots, A_{r-1}$. Suppose that $E_0, \ldots, E_{r-1}$ are the matrices of the orthogonal projections onto the eigenspaces. Recall that

$$A_i = \sum_{j=0}^{r-1} (P)_{ij} E_j,$$

and

$$E_i = n^{-1} \sum_{j=0}^{r-1} (Q)_{ij} A_j.$$

The image of E_j is the jth eigenspace, W_j, and so affords a character of the automorphism group – indeed, the jth irreducible constituent of the permutation character, if the scheme arises from a group. We now see how to calculate values of this character.

Let g be any automorphism, and $P(g)$ the corresponding permutation matrix. Thus, $P(g)$ commutes with all the A_i, and with all the E_j. For $i = 0, \ldots, r-1$, let $\alpha_i(g)$ be the number of points $x \in \Omega$ for which $(x, xg) \in R_i$. Then

$$\alpha_i(g) = \mathrm{Trace}(P(g)A_i).$$

Also, $P(g)E_j = E_jP(g)$, restricted to W_j, is the linear transformation of W_j induced by g, while this matrix annihilates all W_k for $k \neq j$. So the character value $\chi_j(g)$ is given by

$$\chi_j(g) = \mathrm{Trace}(P(g)E_j).$$

Hence we have

$$\chi_i(g) = n^{-1} \sum_{j=0}^{r-1} (Q)_{ij} \alpha_j(g).$$

So, knowledge of the intersection matrices and the numbers $\alpha_i(g)$ enables the character values to be calculated. Note that every character value must be an algebraic integer (since it is a sum of roots of unity); in particular, if the value is rational, then it is an integer.

This gives a necessary condition for the existence of certain automorphisms. The method has not been widely applied, since knowledge of the numbers $\alpha_i(g)$ is not easy to come by. One of the most notable successes of the method is the following result of Graham Higman (unpublished), promised in the last section.

Theorem 3.13 *There is no vertex-transitive Moore graph of diameter* 2 *and valency* 57.

Proof. The proof involves analysing in detail a possible automorphism g of order 2 of such a Moore graph.

Step 1. The fixed points of g form either a star (a vertex and some of its neighbours) or a Moore subgraph.

This follows because, if two non-adjacent vertices are fixed by g, then so is their unique common neighbour. (See Exercise 3.17.)

Step 2. If g interchanges two adjacent vertices, then it fixes 56 vertices altogether.

For let g interchange a and b. Let A be the set of neighbours of a other than b, and B the set of neighbours of b other than a. Then g interchanges A and B. Any further vertex is joined to one neighbour of a (in A) and to one neighbour of b (in B), and so can be labelled with an element of $A \times B$; every label occurs just once. Now a vertex is fixed if and only if its label is (a', b'), where g interchanges a' and b'; there are 56 such vertices.

Step 3. g fixes 56 or 58 vertices, forming a star.

By Step 2, we may suppose that g does not interchange two adjacent vertices. Then, if $ag \neq a$, the vertices a and ag have a unique common neighbour, which is the unique neighbour of a fixed by g. So each non-fixed vertex is joined to a unique fixed vertex. Counting shows that this is impossible unless g fixes a 58-point star. For example, if the fixed vertices formed a 50-point Moore graph, each fixed vertex would have $57 - 7 = 50$ non-fixed neighbours; but $50 \cdot 50 \neq 3250 - 50$.

Step 4. g fixes 56 vertices.

We now know enough about g in the other case to eliminate it by character theory. From our analysis of strongly regular graphs, we find that the P and Q matrices are

$$P = \begin{pmatrix} 1 & 1 & 1 \\ 57 & -8 & 7 \\ 3192 & 7 & -8 \end{pmatrix}, \quad Q = \begin{pmatrix} 1 & 1 & 1 \\ 1520 & -640/3 & 10/3 \\ 1729 & 637/3 & -13/3 \end{pmatrix}.$$

So the projection onto the eigenspace of dimension 1520 is

$$\frac{1}{3250}(1520A_0 - (640/3)A_1 + (10/3)A_2).$$

Now the element g has $\alpha_0(g) = 58$, $\alpha_1(g) = 0$, $\alpha_2(g) = 3192$; so the character value is

$$\chi_1(g) = \frac{1}{3250}(1520 \cdot 58 + 3192 \cdot 10/3) = \frac{152}{5},$$

a contradiction.

Step 5. Conclusion of the proof. Suppose that Γ has a vertex-transitive automorphism group G. Then 3250 divides $|G|$, so $|G|$ is even, and G contains an involution g. We know that g fixes 56 vertices, so $|G_\alpha|$ is even, and $|G|$ is divisible by 4. Let H be the intersection of G with the alternating group A_{3250}. Then $|H|$ is even, so H contains an involution. But, finally, we have a contradiction, since we have shown that involutions have 56 fixed points and $(3250 - 56)/2 = 1597$ transpositions, and so are odd permutations.

Remark. It does not seem so easy to eliminate by this method the possibility of an involution with 56 fixed points. For such an element, we have $\alpha_0(g) = 56$, $\alpha_1(g) = 112$, $\alpha_2(g) = 3082$, from which we compute the character value to be 22.

3.8 Valency bounds

We saw in Chapter 1 that a permutation group is primitive if and only if every orbital graph is connected. Now the diameter of an orbital graph does not exceed the rank, since pairs of points at different distances lie in different orbits. In this way, the degree of a primitive group can be bounded if the rank and a subdegree are known.

If G is primitive and has a suborbit of size 1 other than $\{\alpha\}$, then G is regular and cyclic of prime degree. (For let $\{\beta\}$ be this suborbit, and let $g \in G$ map α to β. Then $g^{-1}G_\alpha g = G_\beta = G_\alpha$, so $N_G(G_\alpha) \geq \langle G_\alpha, g\rangle$. Now G_α is a maximal subgroup of G, as G is primitive, and $g \notin G_\alpha$; so $\langle G_\alpha, g\rangle = G$, and $G_\alpha \trianglelefteq G$. Then G_α fixes every point of Ω, that is, $G_\alpha = 1$. So we may consider subdegrees greater than 1.

Theorem 3.14 *Let G be a primitive permutation group on Ω. Suppose that G has rank r, and that G_α has an orbit of size $k > 1$. Then*

$$|\Omega| \leq \begin{cases} 2r-1 & \text{if } k = 2, \\ 1 + \frac{k((k-1)^{r-1}-1)}{k-2} & \text{if } k > 2. \end{cases}$$

Proof. Suppose first that the orbital graph corresponding to the given suborbit is undirected (that is, the orbital is self-paired). This graph has diameter at most $r-1$ and valency k. Now, in such a graph, any vertex at distance i from α has at most $k-1$ neighbours at distance $i+1$ from α, since at least one neighbour is at distance $i-1$ from α. By induction, the number of points at distance i from α is at most $k(k-1)^{i-1}$ for $i > 0$. Summing a geometric progression, we find $k((k-1)^{r-1}-1)/(k-2)$ (or $2(r-1)$, if $k = 2$) as an upper bound for the number of vertices different from α, as claimed.

If the orbital graph is directed, we count instead the vertices which can be reached from α by *alternating paths* of length i (where the edges are directed alternately away from and towards α.

When does equality hold? If $k = 2$, then every non-trivial suborbit has size 2. Suppose first that the orbital graph is undirected. Then it is a cycle, and G is a dihedral group, necessarily of prime degree (by primitivity). Now we will show that this holds in any case.

Suppose that β lies in a G_α-orbit of size 2. Since a subgroup of index 2 is normal, we have $G_{\alpha\beta} \trianglelefteq G_\alpha$, and similarly $G_{\alpha\beta} \trianglelefteq G_\beta$. By primitivity, G_α is maximal in G, and so $N_G(G_{\alpha\beta}) = G$, whence $G_{\alpha\beta} = 1$. Then $|G_\alpha| = 2$, and G has a self-paired orbital with subdegree 2; so G is dihedral, by the preceding paragraph. It follows that all its orbitals are self-paired.

So we may suppose that $k > 2$.

We note that equality is impossible if the orbital graph is directed, since at least one other suborbit has size k, and $k < k(k-1)^{i-1}$ for $i > 1$ when $k > 2$. Now the argument shows that any point at distance i from α is joined to a unique point at distance $i-1$ from α, and that all its other neighbours are at distance $i+1$ from α if $i < r-1$. So the graph is distance-regular, with $c_i = 1$ for all i, and $a_i = 0, b_i = k-1$ for $i < d$, $a_d = k-1$ (where $d = r-1$ is the diameter).

Such a graph is a *Moore graph*, in the following sense: a *Moore graph* of diameter d is a regular connected graph of diameter d which has no cycles of length less than $2d+1$. We assume that $d \geq 2$: according to this definition, a Moore graph of diameter 1 would be a complete graph.

A Moore graph of valency 2 and diameter d is a cycle of length $2d+1$. We saw earlier that a Moore graph of diameter 2 must have valency 2, 3, 7 or (possibly) 57. It was shown by Hoffman and Singleton [94] for $d = 3$, and by Bannai and Ito [13] and Damerell [58] in general, that there are no further Moore graphs:

Theorem 3.15 *There is no Moore graph with diameter and valency greater than* 2.

Hence we can say exactly when equality holds in the valency bound:

Theorem 3.16 *Equality in Theorem 3.14 holds only in the following cases:*

(a) $r = 2$, *G is 2-transitive;*

(b) $k = 2$, *G is dihedral of prime degree;*

(c) $r = 3$, $k = 3$, *G is S_5 or A_5 (degree* 10*);*

(d) $r = 3$, $k = 7$, *G is* $\mathrm{P\Sigma U}(3,5^2)$ *or* $\mathrm{PSU}(3,5^2)$ *(degree* 50*).*

3.9 Distance-transitive graphs

A distance-transitive graph of valency 2 is simply a cycle; there are infinitely many such cycles up to isomorphism, one of each length greater than 2. The position is different for larger valency, however. One of the first major results about distance-transitive graphs was the theorem of Biggs and Smith [18]:

Theorem 3.17 *There are exactly twelve finite distance-transitive graphs with valency* 3*, up to isomorphism.*

Briefly, the analysis goes as follows. We use a remarkable theorem proved by Tutte [177] in 1947:

Theorem 3.18 *Let Γ be a finite connected graph of valency* 3 *whose automorphism group G is transitive on vertices and edges. Then the stabiliser of a vertex in G has order $3 \cdot 2^a$, where $a \leq 4$.*

Since the stabiliser of a vertex x acts transitively on the set of vertices at distance d from x, this set has size at most $3 \cdot 2^4 = 48$. Thus, the graph cannot 'fan out' like the trivalent tree for more than five stages, but must begin to 'close up'. From this fact, Biggs and Smith show by graph-theoretic arguments that the graph can have diameter at most 15. (Loosely: by at most three times as far as the point where it begins to close up, it has completely closed.) This already demonstrates that there are only finitely many distance-transitive graphs of valency 3.

To determine them completely, Biggs and Smith used a computer to list all the possible intersection arrays for such graphs, and to test them for feasibility by computing the multiplicities of their eigenvalues. For those arrays which pass all possible tests, it is necessary to decide whether a distance-transitive graph exists, and if it is unique up to isomorphism.

Subsequently, Smith and other authors gave similar determinations of the distance-regular graphs of valency $4, 5, \ldots$. In each case, the essential steps in the argument run as follows:

(a) Bound the order of the vertex stabiliser.

(b) Bound the diameter.

(c) Analyse the finitely many possible configurations.

I shall refer to this as *Smith's program.* The first step is closely related to the *Sims Conjecture*:

Conjecture. Let G be a finite primitive permutation group on Ω, and suppose that G_α has an orbit of length k (other than $\{\alpha\}$). Then $|G_\alpha|$ is bounded by a function of k.

Of course, the automorphism group of a distance-transitive graph need not be primitive, so the conjecture does not immediately apply. However, Smith showed that there is a very strong reduction to the primitive case. Essentially, there are only two ways in which a distance-transitive graph can be imprimitive, as follows. Let Γ be a distance-transitive graph with diameter d.

- Γ is *bipartite* if its vertices can be partitioned into two sets B_1 and B_2 (called *bipartite blocks*) so that every edge of Γ has one vertex in each block. If this is the case, then the bipartite blocks are blocks of imprimitivity, and two vertices lie in the same block if and only if their distance is even. Moreover, if we define $\Gamma^{\{2\}}$ to be the 'distance-2 graph', in which two vertices are adjacent if and only if they lie at distance 2 in Γ, then $\Gamma^{\{2\}}$ has two connected components $\Gamma_1^{\{2\}}$ and $\Gamma_2^{\{2\}}$, one on each bipartite block; these two components are isomorphic and each is distance-transitive, with diameter $\lfloor d/2 \rfloor$. We say that Γ is a *bipartite double* of $\Gamma_1^{\{2\}}$.

- Γ is *antipodal* if the relation of being equal or at distance d is an equivalence relation on the vertex set. If this occurs, the equivalence classes are called *antipodal classes*, and they are blocks of imprimitivity for $\mathrm{Aut}(\Gamma)$. The graph $\overline{\Gamma}$, whose vertices are the antipodal classes, two vertices adjacent if they contain adjacent vertices of Γ, is the *antipodal quotient* of Γ; it is distance-transitive, with diameter $\lfloor d/2 \rfloor$. We say that Γ is an *antipodal cover* of $\overline{\Gamma}$.

For example, the 3-dimensional cube is bipartite. The distance-2 graph consists of two disjoint complete graphs of size 4. The cube is also antipodal; its antipodal quotient is also isomorphic to K_4. (See Figure 3.3.)

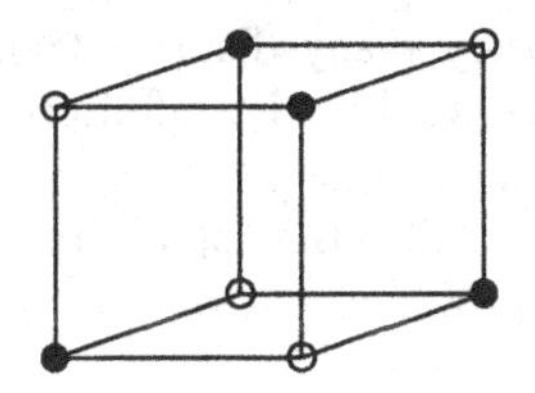
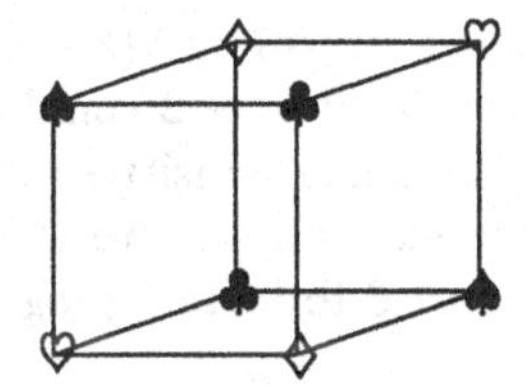

Figure 3.3: Bipartite blocks and antipodal classes

Smith [168] showed that these are the only ways in which a distance-transitive graph of valency greater than 2 can be imprimitive:

Theorem 3.19 *Let Γ be a finite distance-transitive graph of valency greater than 2, whose automorphism group G is imprimitive. Then Γ is either bipartite or antipodal, and a non-trivial block of imprimitivity is either a bipartite block or an antipodal class.*

Proof. Let $\equiv$ be a non-trivial congruence on the vertex set of Γ. By distance-transitivity, there is a subset I of $\{0, 1, \ldots, d\}$, where d is the diameter of Γ, such that $x \equiv y$ if and only if $d(x, y) \in I$. Now $0 \in I$, by reflexivity, and $1 \notin I$ (since if $1 \in I$ then $I = \{0, 1, \ldots, d\}$ by connectedness). If $I = \{0, d\}$, then Γ is antipodal, and blocks are antipodal classes; so assume not. Let e be the smallest positive number in I. By considering a geodesic path of length d, we see that I is closed under subtraction, and also under addition if the sum does not exceed d. Hence I consists of all multiples of e not exceeding d. As a consequence, if $e = 2$, then Γ is bipartite, and the blocks of imprimitivity are the two bipartite blocks. So we may assume that $3 \le e \le d - 1$.

Now $a_e = 0$; for, if two vertices u, v of $\Gamma_e(x)$ were adjacent, then $u \equiv x \equiv v$, a contradiction (since $e > 1$). Also $b_{e-1} = 1$; for, if $b_{e-1} > 1$, there are two vertices $u, v \in \Gamma_e(x)$ joined to the same vertex $y \in \Gamma_{e-1}(x)$; but then $d(u,v) \leq 2$ and $u \equiv x \equiv v$, a contradiction. Since $b_{e-1} \geq b_e$ by Theorem 3.7, we have $b_e = 1$. Similarly, $c_{e+1} = 1$, and so $c_e = 1$. But then

$$k = c_e + a_e + b_e = 1 + 0 + 1 = 2,$$

contrary to the assumption that $k \geq 3$.

Now one can show the following. Assume that Γ is distance-transitive, with valency $k > 2$. If Γ is bipartite, then the components of its distance-2 graph are not bipartite; and, if Γ is antipodal, then its antipodal quotient is not antipodal. So, in at most two such reductions, we reach a graph with primitive automorphism group. Moreover, a graph and its antipodal cover have the same valency; and a bipartite double of a graph of valency k must have valency $kc_1/b_2 \leq k(k-1)$. So, if infinitely many distance-transitive graphs of valency $k > 2$ could exist, then there would exist infinitely many primitive distance-transitive graphs of some valency $k' > 2$.

In the next chapter, we will see how the Classification of Finite Simple Groups has led to further progress on Smith's program.

3.10 Multiplicity bounds

Just as the degree of a primitive permutation group is bounded by a function of its rank and one non-trivial subdegree, so it is bounded by a function of its rank and one non-trivial character degree. However, the techniques are quite different. I will prove just a special case.

Theorem 3.20 *Let G be a primitive permutation group on Ω, having rank r. Suppose that the permutation character of G contains a non-principal real irreducible constituent with degree f and multiplicity* 1. *Then*

$$|\Omega| \leq \binom{f+r-2}{r-1} + \binom{f+r-3}{r-2}.$$

Proof. Let E be the matrix of orthogonal projection onto the subspace W of dimension f affording the given character χ. Since E commutes with the permutation matrices, the image of the standard basis consists of vectors v_α, for each $\alpha \in \Omega$, which are permuted by the representation affording χ as Ω is permuted by G.

The vectors v_α are all distinct. For the relation on Ω, defined by $\alpha \equiv \beta$ if and only if $v_\alpha = v_\beta$, is a congruence; it is not the universal relation, since χ is non-principal; since G is primitive, it is the relation of equality. Also, G is represented on W by real orthogonal matrices, so inner products between the vectors v_α are preserved. In particular, all these vectors have the same length; by re-scaling, we may assume that they lie on the unit sphere. Also, since G has rank r, there can be at most $r-1$ different values of the inner product of two distinct vectors.

The result follows from a theorem of Delsarte, Goethals and Seidel [60]:

Theorem 3.21 *Let A be a set of points on the unit sphere in $\mathbb{R}^f$, such that $a.b$ takes at most s distinct values for $a, b \in A$, $a \neq b$. Then*

$$|A| \leq \binom{f+s-1}{s} + \binom{f+s-2}{s-1}.$$

Proof. Let S be the set of values of the inner product $a.b$ for $a, b \in A$, $a \neq b$. For $a \in A$, let

$$f_a(x) = \prod_{\sigma \in S} \frac{x.a - \sigma}{1 - \sigma}.$$

Then $f_a(b) = \delta_{ab}$ (the Kronecker delta), so the functions f_a are linearly independent.

Now f_a is a polynomial function of degree s, so $|A|$ does not exceed the dimension of the space of polynomial functions of degree s on the unit sphere. For each i, the number of monomials of degree i that can be formed from f variables is $\binom{f+i-1}{i}$. So the sum of these binomial coefficients for $i \leq s$ is an upper bound.

But we can do better. On the unit sphere, $x.x = 1$; so, from a homogeneous polynomial of degree i, we obtain one of degree $i + 2k$ with the same value by multiplying by $(x.x)^k$. So our polynomial of degree s can be written as a sum of homogeneous polynomials of degrees s and $s-1$. So the dimension is bounded by $\binom{f+s-1}{s} + \binom{f+s-2}{s-1}$, as required.

Many further necessary conditions are known. See, for example, the books [14, 25]. We mention just one type, the so-called *Krein conditions.* They depend on the following result. The *Hadamard product* of two matrices $A = (a_{ij})$ and $B = (b_{ij})$ is the matrix $A \circ B$ whose (i, j) entry is $a_{ij}b_{ij}$; that is, obtained by coordinatewise multiplication. Now let $(\Omega, \mathcal{R})$ be an association scheme. Since each basis matrix has entries 0 and 1, and the ones occur in disjoint positions, the basis matrices satisfy

$$A_i \circ A_j = \delta_{ij} A_i,$$

where δ_{ij} is the Kronecker delta. So the basis algebra is closed under Hadamard multiplication. In particular, the idempotents $E_0, \ldots, E_{r-1}$ satisfy

$$E_i \circ E_j = \sum_{k=0}^{r-1} q_{ij}^k E_k$$

for some complex numbers q_{ij}^k, called the *Krein parameters* of the scheme.

Theorem 3.22 *The Krein parameters of a symmetric association scheme are real numbers in the interval* $[0, 1]$.

Proof. The matrices E_i are positive semidefinite, since their eigenvalues are all 0 or 1. Now the eigenvalues of the tensor or Kronecker product $E_i \otimes E_j$ are obtained by multiplying those of E_i and E_j, so are again 0 and 1. The Hadamard product is a principal submatrix of the tensor product, so its eigenvalues lie in any interval including all the eigenvalues of the tensor product, which we can take to be $[0, 1]$. But the formula

$$E_i \circ E_j = \sum_{k=0}^{r-1} q_{ij}^k E_k$$

shows that the numbers q_{ij}^k, for $k = 0, \ldots, r-1$, are the eigenvalues of $E_i \circ E_j$. (The displayed equation is the spectral decomposition of this matrix.)

3.11 Duality

Delsarte [59] developed a theory of duality for association schemes, which applies (in particular) to all schemes which admit a regular abelian group of automorphisms. We will describe here only the group case. First, we need to prove a result of Brauer. We recall from Chapter 2 the special case of *Block's Lemma* which applies to square incidence matrices. Let G act on two sets Ω_1 and Ω_2 of the same size. Suppose that there is an invertible matrix M such that, if P_1 and P_2 are the permutation representations of G on Ω_1 and Ω_2, then

$$P_1(g)M = MP_2(g)$$

for all $g \in G$. Then G has the same permutation character on Ω_1 and Ω_2. This is immediate from the more general form of Block's Lemma, but is easily proved directly:

$$\mathrm{Trace}(P_1(g)) = \mathrm{Trace}(M^{-1}P_2(g)M) = \mathrm{Trace}(P_2(g)).$$

Now let A be a group, and let G be a group of automorphisms of A. Then G induces a permutation of the conjugacy classes of A, since if two elements of A are conjugate, then so are their images. Also, G acts on the irreducible characters of A, by the rule

$$\chi^g(a) = \chi(a^{g^{-1}}),$$

where we have written $a^{g^{-1}}$ for the image of a under the automorphism g^{-1}. (The inverse is, as usual, necessary to ensure that we have an action; it does not affect the computation of orbits.) Brauer's Lemma asserts:

Theorem 3.23 *Any group of automorphisms of a group A has equally many orbits on conjugacy classes and on irreducible characters of A.*

This follows on applying Block's Lemma, using the character table of A as the matrix M.

Let G be a permutation group on Ω, with a regular normal abelian subgroup A. As in Section 1.6, we can identify Ω with A so that A acts by right multiplication and G_1 acts as a group of automorphisms. Since A is its own centraliser, any basis matrix for G is a sum of permutation matrices corresponding to elements of A; so the basis matrices commute, and the centraliser algebra is commutative. Thus, the coherent configuration associated with G is a commutative association scheme. The rank of G is the number of orbits of G^* on A. Moreover, G is a semidirect product $A \rtimes G_1$.

Let A^* be the dual group of A (see Section 2.9). Then A^* is an abelian group isomorphic to A. Moreover, G_1 acts as a group of automorphisms of A^*, having the same number of orbits as on A. Thus, $G^* = A^* \rtimes G_1$ is a permutation group on a set Ω^*, having an abelian regular normal subgroup A^* isomorphic to A, and with the same point stabiliser as in G. The group G^* and its association scheme are said to be *duals* of G and its scheme.

Many properties are shared by the scheme and its dual, which we now state.

Theorem 3.24 *Let G be a permutation group with an abelian regular normal subgroup A, and let G^* be its dual.*

(a) G and G^* have the same rank.

(b) G is primitive if and only if G^* is primitive.

(c) The valencies of G are the multiplicities of G^* and vice versa.

(d) The P and Q matrices of G are the Q and P matrices of G^* respectively.

(e) The Krein parameters of G, multiplied by $n = |A|$, are the intersection numbers of G^*, and vice versa.

Proof. We have shown (a) already; we will sketch the proofs of (b) and (c). The blocks of imprimitivity for G containing the identity of A are the G_1-invariant subgroups of A. If B is such a subgroup, then its annihilator $B^\dagger$ is a G_1^*-invariant subgroup of A^*, and conversely. (Note that the number of blocks and the size of a block are exchanged in passing from G to G^*.)

The basis matrices of G are just the sums, over the G_1-orbits, of the basis matrices of A; and these last are just the elements of A, regarded as permutation matrices. Thus, each character of A gives rise to a character of the basis algebra for G. Now characters which are in the same G_1^*-orbit must take the same value on each basis matrix. Since the numbers of orbits are equal, we see that two characters lie in the same G_1^*-orbit if and only if they give rise to the same character of the basis algebra of G. Thus, the multiplicities of G are just the orbit sizes of G_1^* on A^*, which are the valencies of G^*. The fact that the valencies of G are the multiplicities of G^* is proved by dualising, since G^{**} is naturally isomorphic to G.

Example. The 4-dimensional cube admits the group $(C_2^4) \rtimes S_4$. It is isomorphic to its dual: for there is a natural inner product on the 4-dimensional GF(2)-vector space which is preserved by G. However, C_2^4 has an invariant subgroup C_2^3, corresponding to those vectors for which the sum of the coordinates is zero. The group $(C_2^3) \rtimes S_4$ has rank 3 and subdegrees $1, 1, 6$: it is imprimitive, since the constant vectors form an invariant 1-dimensional subspace. So in the dual group, there is an invariant 2-dimensional subspace, and the subdegrees are $1, 3, 4$ (which are the multiplicities for the original).

It is harder to find primitive groups which are not isomorphic to their duals; but there are examples associated with semidirect products of abelian groups related to Golay codes by Mathieu groups. See, for example, Exercise 3.15.

3.12 Wielandt's Theorem

As an example of how very specific information can be obtained from the knowledge of eigenvalue multiplicities, we now discuss a theorem due to Wielandt [183].

Theorem 3.25 *Let p be prime, and let G be a primitive permutation group of degree $2p$ which is not 2-transitive. Then $p = 2s^2 + 2s + 1$ for some positive integer s, and G has rank 3 and subdegrees $1, s(2s+1), (s+1)(2s+1)$.*

Proof. In the first part of the argument, following Burnside's proof of his theorem on groups of prime degree (see Section 2.1), Wielandt shows that G does indeed have rank 3, and that the irreducible constituents of the permutation character have degrees $1, p-1, p$. Thus, G is a group of automorphisms of a strongly regular graph. The graph falls under Case II, since the number of vertices is even. Take its valency to be k, and the other eigenvalues of its adjacency matrix (which are integers in this case) to be r and s. Replacing the graph by its complement if necessary, we may assume that $k < p$. Now, if A is the adjacency matrix of the graph, we have

$$\begin{aligned} 0 &= \mathrm{Trace}(A) &= k + (p-1)r + ps, \\ 2pk &= \mathrm{Trace}(A^2) &= k^2 + (p-1)r^2 + ps^2. \end{aligned}$$

Thus $r \equiv k \pmod{p}$. Since $k < p$ and $r \neq k$, we have $r = k - p$. Then calculation gives the result: we find $p - k = s + 1$, and $p = (p-k)^2 + s^2 = 2s^2 + 2s + 1$.

Remark. The groups S_5 and A_5 both act primitively on the ten 2-element subsets of $\{1, \ldots, 5\}$. In the terminology of Wielandt's theorem, we have $s = 1$; the subdegrees are $1, 3, 6$. The strongly regular graph in the proof is the Petersen graph in this case.

Using the Classification of Finite Simple Groups, which we discuss in the next chapter, it has been shown that there are no further examples.

3.13 Exercises

3.1. If a coherent configuration is associated with the right regular representation of a group G, show that its basis algebra is spanned by the elements of the left regular representation of G.

3.2. Prove that, if the product of two symmetric matrices A and B is symmetric, then $AB = BA$.

3.3. Prove Theorem 3.7(b).

3.4. Show that the coherent configuration of a primitive rank 3 permutation group of even order is obtained from a distance-transitive graph of diameter 2.

3.5. Prove the assertions made about the reduction to primitivity for distance-transitive graphs immediately after Smith's Theorem 3.19.

3.6. Show that two of the graphs in the Johnson scheme $J(2k+1, k)$ are distance-transitive: the Johnson graph (in which two k-sets are adjacent if they meet in $k-1$ points) and the so-called *odd graph* O_{k+1} (in which two k-sets are adjacent if and only if they are disjoint).

3.7. Show that the Hamming graphs $H(n, 2)$ and the Johnson graphs $J(2k, k)$ are antipodal, and describe their antipodal quotients. Show also that the Hamming graphs $H(n, 2)$ are bipartite, and describe their distance-2 graphs.

3.8. (a) Let G be a group, all of whose irreducible characters are real-valued (for example, the symmetric group, see Section 2.10). Show that a permutation character of G is multiplicity-free if and only if all the orbitals are self-paired.

(b) Prove that, if $n > 15$, then the permutation character of S_n on partitions into l sets of size k (where $kl = n$ and $k, l > 1$) is multiplicity-free if and only if k or l is 2.

3.9. A *Latin square* is an $n \times n$ array with entries taken from the set $N = \{1, 2, \ldots, n\}$, having the property that each symbol from N occurs exactly once in each row and column of the array.

(a) Given a Latin square L, define three relations R_0, R_1, R_2 on the set Ω of n^2 cells of the array as follows:

- R_0 is the relation of equality;
- two cells are in the relation R_1 if they lie in the same row, or in the same column, or contain the same symbol (in the square L);
- two cells are in the relation R_2 if neither R_0 nor R_1 holds for them.

Prove that $(\Omega, \{R_0, R_1, R_2\})$ is a coherent configuration, whose intersection numbers depend only on n, not on the particular Latin square used.

(b) Let L_1 and L_2 be the Cayley tables of the two groups of order $n = 4$. Prove that the configurations obtained from L_1 and L_2 are not isomorphic, even though their intersection algebras are identical.

Remark. The graph with adjacency relation R_1 is called a *Latin square graph.* It is strongly regular.

3.10. Show that the conjugacy class scheme of a group G (the coherent configuration associated with the diagonal group

$$G^* = \{x \mapsto g^{-1}xh : g, h \in G\}$$

acting on G) is a commutative association scheme. How is its Q matrix related to the character table of G?

3.11. Prove that the complement of a strongly regular graph is strongly regular. Prove also that a strongly regular graph has the same intersection numbers as its complement if and only if it falls under Case I of our analysis.

3.12. (a) Prove that there is a unique Moore graph of diameter 2 and valency 3, up to isomorphism.

(b) Form a graph whose vertices are the 2-subsets of $\{1, \ldots, 5\}$, two vertices adjacent if and only if they are disjoint. (This is the *odd graph* O_3.) Prove that it is a Moore graph with diameter 2 and valency 3.

(c) Prove that the full automorphism group of the graph constructed in (b) is S_5.

(d) Write down the P and Q matrices of the Petersen graph. Hence decompose the permutation character of S_5 on the vertices of the graph into irreducible constituents.

3.13. By computer, or otherwise, verify that the following construction gives a Moore graph of valency 7 on 50 vertices with a transitive rank 3 automorphism group.

Take $G = S_6$. Then G has two transitive actions of degree 6, the first being the natural one on $A = \{1, \ldots, 6\}$, and the second on X, the set of cosets of $\mathrm{PGL}(2,5)$. Now take the vertex set of the graph to be $\{\alpha, \chi\} \cup A \cup X \cup (A \times X)$, with the natural action of S_6 (fixing α and χ). The edges are as follows:

- α is joined to χ;
- α is joined to all points of A;
- χ is joined to all points of X;
- $(a, x) \in A \times X$ is joined to $a \in A$ and to $x \in X$;
- take also the edges of the unique orbital graph in $A \times X$ which is connected and has valency 5.

3.14. (a) Prove that a regular graph in which any two vertices have a unique common neighbour is a triangle.

(b) Prove that a non-regular graph with the above property has a vertex which is joined to all others.

Remark. The content of this exercise is the *Friendship Theorem* of Erdős, Rényi and Sós [67], according to which, in a finite society in which any two members have a unique common friend, there is somebody who is everybody else's friend.

3.15. Show that the even-weight subcode of the binary Golay code is an 11-dimensional subspace of the vector space $\mathrm{GF}(2)^{23}$ admitting an action of the Mathieu group M_{23}, so that the semidirect product is primitive and has rank 4 and subdegrees 1, 253, 506, 1288. Show that its dual has subdegrees 1, 23, 253, 1771. (See [43] for information about Golay codes.)

3.16. Let G be a permutation group which has a regular normal subgroup A which is an elementary abelian p-group. Prove that Frame's constant for G is a power of p.

3.17. Let Γ be a graph with no triangles, in which any two non-adjacent vertices have a unique common neighbour. Prove that either

(a) Γ is a star, consisting of one vertex adjacent to a set of pairwise non-adjacent vertices; or

(b) Γ is regular, and is a Moore graph of diameter 2.

3.18. Verify the character value stated in the text for an involution acting on a Moore graph of valency 57 with 56 fixed points.

3.19. Prove that the number of monomials of degree i that can be formed from f variables is $\binom{f+i-1}{i}$.

3.20. A *principal submatrix* of a matrix A is the matrix formed by deleting a set of rows and the corresponding columns from A. Prove that a principal submatrix B of a positive semidefinite symmetric real matrix is positive semidefinite symmetric. Hence show that, if A is symmetric and λ, μ are its least and greatest eigenvalues, then all eigenvalues of B lie in the interval $[\lambda, \mu]$.

3.21. Show that the integrality conditions for the eigenvalue multiplicities of a putative strongly regular graph with $k = 9$, $\lambda = 0$, and $\mu = 4$ are satisfied. Use either the multiplicity bound or the Krein bound (or both) to show that there is no such graph.

3.22. Let Γ be a strongly regular graph on $2m$ vertices in which the eigenvalue multiplicities are $1, m-1, m$. Show that either

(a) Γ or its complement is a disjoint union of edges; or

(b) $m = 2a^2 + 2a + 1$, and γ or its complement has parameters

$$k = a(2a+1), \quad \lambda = (a-1)(a+1), \quad \mu = a^2.$$

3.23. Let Γ be a Moore graph of valency 3 or 7. Define a new graph Δ whose vertices are the edges of Γ, two vertices adjacent if and only if as edges they are disjoint but there is an edge which meets both. Show that Δ is strongly regular.

3.24. Let Γ be a strongly regular graph of valency k. Suppose that Γ has a 1-factor (a set of pairwise disjoint edges which covers all the vertices) having the property that the permutation which interchanges the ends of each edge in the factor is an automorphism of Γ. Let r be an eigenvalue of Γ with multiplicity f. Prove that k divides rf.

CHAPTER 4

The O'Nan–Scott Theorem

4.1 Introduction

We have already seen two reductions for a finite permutation group:

(a) An intransitive group is a subcartesian product of its transitive constituents.

(b) A transitive but imprimitive group is contained in the iterated wreath product of its primitive components.

These enable many questions about arbitrary groups to be reduced to the case of a primitive group and hopefully solved there. In this section, we make one further reduction, and then attempt to describe the 'basic' building blocks we have reached.

The *socle* of a finite group G is the product of the minimal normal subgroups of G. (The original meaning of the word is 'the base on which a statue stands'.) The socle is a normal subgroup whose structure can be described: it is a direct product of finite simple groups (but, in general, not all of these simple groups are isomorphic). We will see that, for a primitive group G, either the socle is a product of isomorphic finite simple groups in a known permutation action, or G itself is almost simple. In many practical problems, we can then appeal to the *Classification of Finite Simple Groups* (CFSG) to reach a conclusion in the last case.

A version of this theorem (though without much of the detail) which is sufficient for some of the applications appears in Jordan's *Traité des Substitutions* [103]. But since the use of the theorem depends on knowledge of the finite simple groups which was not available for 120 years after Jordan, his result was forgotten. The modern version was proved independently by Michael O'Nan and Leonard Scott and announced at the Santa Cruz Conference on Finite Groups in 1979, just before the CFSG was announced. Both papers appeared in the preliminary version of the conference proceedings, but O'Nan's paper disappeared from the final version.

4.2 Precursors

Several earlier results on normal (or minimal normal) subgroups have become 'classical'.

Theorem 4.1 (Jordan's Theorem) *Let G be k-transitive, but not S_k, with $k > 1$. Then a non-trivial normal subgroup N of G is $(k-1)$-transitive, except possibly in the case $k = 3$, when N may be an elementary abelian 2-group.*

Proof. The proof is by induction on k. For $k = 2$, the 2-transitive group G is primitive, and so its normal subgroup N is transitive, as required. So assume that $k \geq 3$, and that the result is true with $k-1$ replacing k.

Choose $\alpha \in \Omega$. Then N_α is a normal subgroup of the $(k-1)$-transitive group G_α (but possibly trivial). By the induction hypothesis, one of three possibilities occurs:

- $N_\alpha = 1$. Then N is regular. In our analysis of regular normal subgroups in Section 1.7, we showed that N is an elementary abelian 2-group, and that G is not 4-transitive.
- N_α is $(k-2)$-transitive. Since N is transitive, it is $(k-1)$-transitive.
- $k = 4$, and N_α is a regular elementary abelian 2-group. Then N is sharply 2-transitive. Let $n \in N$ interchange 1 and 2. Since N is sharply 2-transitive, $n^2 = 1$. Suppose that n also interchanges 3 and 4. Using the 4-transitivity of G and the fact that $n \geq 5$, choose $g \in G$ to fix 1, 2 and 3, and to map 4 to 5. Then $g^{-1}ng \in N$ has cycle structure $(1\ 2)(3\ 5)\cdots$, which is impossible, since N is sharply 2-transitive and cannot contain two different elements interchanging 1 and 2.

This theorem also holds without modification for infinite groups.

Wagner [178] found an elegant proof of a strengthening of Jordan's Theorem. He showed:

Theorem 4.2 *A non-trivial normal subgroup of a 3-transitive permutation group of odd degree greater than 3 is 3-transitive.*

Proof. Let G be 3-transitive of odd degree $n > 3$, and N a non-trivial normal subgroup of G. Then N is 2-transitive, by Jordan's Theorem; and $N_{\{\alpha,\beta\}}$ (the stabiliser of the unordered pair) has k orbits of length l on $\Omega \setminus \{\alpha, \beta\}$, where $kl = n - 2$, so that l is odd. Now $N_{\alpha\beta}$ is a normal subgroup of $N_{\{\alpha,\beta\}}$ with index 2, so each $N_{\{\alpha,\beta\}}$-orbit splits into at most two $N_{\alpha\beta}$-orbits of the same size. Since l is odd, no splitting can occur, and these two subgroups have the same orbits on $\Omega \setminus \{\alpha, \beta\}$.

Let P be a Sylow 2-subgroup of $N_{\{\alpha,\beta\}}$. Then P has at least k fixed points, one in each orbit of $N_{\{\alpha,\beta\}}$. Since $|P|$ is greater than the 2-part of a 2-point stabiliser, it follows that $k = 1$, that is, that N is 3-transitive.

It follows immediately from Wagner's Theorem that the group $\mathrm{PSL}(2, 2^d)$ (which is sharply 3-transitive of degree $2^d + 1$) is simple for $d > 1$. In particular, A_5 is simple. From this, it is easily shown by induction that A_n is simple for all $n \geq 5$ (see Exercise 4.2).

For 2-transitive groups, we have the following:

Theorem 4.3 (Burnside's Theorem) *Let N be a minimal normal subgroup of a finite 2-transitive group G. Then N is either elementary abelian and regular, or simple and primitive.*

Proof. The heart of the proof is the assertion that N is either primitive or regular. So suppose that G is not primitive. Let B be a minimal non-trivial block for N. Since N is normal in G, for any $g \in G$, Bg is also a minimal non-trivial block for N. Since the intersection of blocks is a block, we see that, if $Bg \neq B$, then $|B \cap Bg| \leq 1$.

Let $\mathcal{B}$ be the set $\{Bg : g \in G\}$. Then $\mathcal{B}$ is the set of lines of a *linear space* on Ω: that is, two points lie in a unique 'line' in $\mathcal{B}$. For the above argument shows that two points lie in at most one line, and then the 2-transitivity of G shows that two points lie in exactly one line. Moreover, this linear space admits a *parallelism* (a partition of the lines into 'parallel classes', each of which partitions the points): a parallel class is just the set of blocks in a fixed congruence for N. The group N fixes every parallel class of lines, and hence N_α fixes every line through α.

Now, by 'radiolocation' (see Figure 4.1), we see that $N_{\alpha\beta}$ fixes every point off the line through α and β. (For if γ is any point not on this line, then $N_{\alpha\beta}$ fixes the lines $\alpha\gamma$ and $\beta\gamma$, and hence fixes their intersection, which is γ.) It now easily follows, replacing β by γ in this argument, that $N_{\alpha\beta} = 1$.

By Frobenius' Theorem 2.1, N has a regular normal subgroup M, consisting of the identity and the fixed-point-free elements of N. Then M is a normal subgroup of G: since N was minimal normal, we have $N = M$, and N is regular.

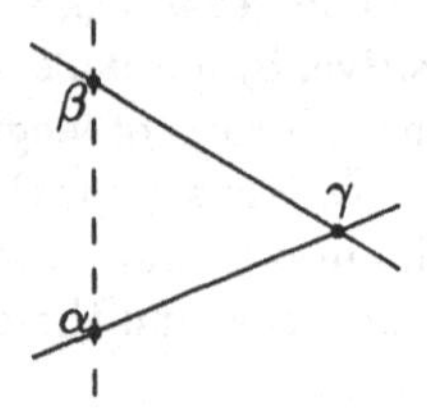

Figure 4.1: Radiolocation

Now to finish the proof: the case where N is regular was considered in Section 1.7, where we showed that it is elementary abelian; and if G is primitive, then it is a product of isomorphic simple groups, and it is not hard to show that there can only be one factor.

This proof fails in the infinite case, in several respects:

- An infinite imprimitive group may not have a minimal non-trivial block.
- Frobenius' Theorem fails in the infinite case.
- The very last step (the proof that if N is regular then it is elementary abelian) fails, as the diagonal group derived from the HNN-construction shows. (Recall that this is a 2-transitive group which is the direct product of two regular normal subgroups.)

We may conclude that, if the normal subgroup N has a minimal block, then it is a Frobenius group; but that is all. I do not know of an example of an infinite 2-transitive group with a Frobenius normal subgroup but no regular normal subgroup.

Our last preliminary result is 'folklore'.

Theorem 4.4 *A primitive permutation group has at most two minimal normal subgroups; if it has two, then they are the left and right regular representations of the same group (and in particular are isomorphic).*

Proof. Any minimal normal subgroup is transitive. If there are two, then they centralise each other, and so they are semiregular (see Exercise 1.5), and hence regular. Then there is no room for a third.

4.3 Product action and basic groups

Recall the *wreath product* $H \text{ Wr } K$ of two permutation groups H (on Γ) and K (on Δ). We defined it as a permutation group on the set $\Gamma \times \Delta$

(regarded as a fibre bundle over Δ with fibres isomorphic to Γ); in this action it is imprimitive, provided that Γ and Δ each have more than one element. We now define another action of the wreath product, the *product action*, on the set of global sections of the fibre bundle. (A *global section* is a subset containing one point from each fibre, see Figure 4.2.)

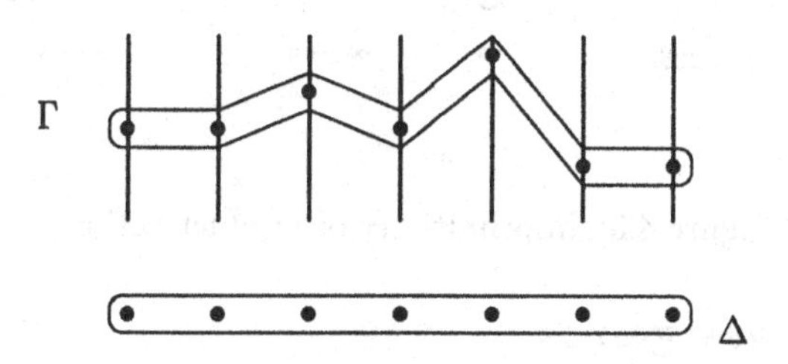

Figure 4.2: A global section

More formally, $H \operatorname{Wr} K$ acts on the set Γ^Δ of all functions from Δ to Γ; the base group B (which is the set of functions from Δ to H) acts coordinatewise, and the top group T (which is isomorphic to K) permutes the arguments of the functions. In detail, for $\phi \in \Gamma^\Delta$, $f \in B = H^\Delta$, we put $(\phi f)(\delta) = \phi(\delta)f(\delta)$, and for $\phi \in \Gamma^\Delta$, $k \in K$, we put $(\phi k)(\delta) = \phi(\delta k^{-1})$ (the inverse is required to obtain an action of K).

Theorem 4.5 *The product action of H Wr K is primitive if and only if H is primitive and not regular on Γ, Δ is finite, and K is transitive on Δ.*

We only consider finite groups here. We illustrate diagrammatically how primitivity fails if one of the other three conditions of the theorem is violated in Figure 4.3.

Note that S_m Wr S_n, in the product action, is the automorphism group of the *hypercubic graph* or *Hamming scheme* $H(n,m)$, as we saw in the last chapter. The vertices of the graph are all the ordered n-tuples taken from an *alphabet* of m symbols: two vertices are adjacent if their *Hamming distance* is equal to 1, that is, if they differ in exactly one coordinate. (For $n = 2$, this is a square grid, as shown.)

We say that the permutation group G is *non-basic* if it is contained in a wreath product with the product action (that is, if it preserves the structure of a Hamming scheme on Ω); it is *basic* otherwise.

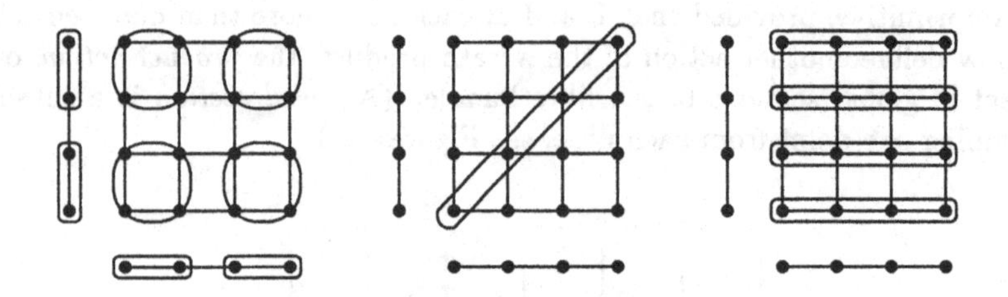

Figure 4.3: Imprimitivity of product action

4.4 Some basic groups

We define two types of permutation groups, affine and diagonal. A group of *affine type* is one satisfying $V \leq G \leq \mathrm{AGL}(V)$ for some vector space V over a finite prime field. We identify V with its additive group, so that $G = V \rtimes G_0$, where $G_0 \leq \mathrm{GL}(V)$, that is, G_0 is a linear group. Now G is necessarily transitive, since it contains the regular normal subgroup V; and

- G is primitive if and only if G_0 is irreducible on V; and
- G is basic if and only if G_0 is *primitive* as a linear group (that is, preserves no direct sum decomposition of V).

(If G_0 preserves a decomposition $V = \bigoplus V_i$, then any vector in V is uniquely expressible as the sum of 'components' from the subspaces V_i; this gives an identification of V with a Hamming scheme.)

Diagonal groups are defined as follows. Let T be a non-abelian simple group, and k a positive integer. The minimal diagonal group is $T^k = T \times \cdots \times T$ (k factors), acting on the cosets of the diagonal subgroup $D = \{(t, t, \ldots, t) : t \in T\}$. For $k = 2$, this is just the diagonal group T^* we have seen several times before. For each coset of D has a unique representative of the form $(1, x)$; and $D(1, x)(g, h) = D(g, xh) = D(1, g^{-1}xh)$. In general there is no such convenient representation, but we can take as a set of coset representatives all those elements of T which have 1 in the first coordinate, and so identify Ω with T^{k-1}.

We can obtain larger groups by adjoining

- permutations of the coordinates,
- automorphisms of T (acting the same way on all coordinates) – but note that inner automorphisms are already included.

So the largest diagonal group has shape $T^k.(\mathrm{Out}(T) \times S_k)$. It is basic. We say that G is of *diagonal type* if $T^k \leq G \leq T^k.(\mathrm{Out}(T) \times S_k)$. In fact, a group G of diagonal type is primitive if and only if the subgroup $\overline{G}$ of S_k induced on the set of direct factors of the socle T^k preserves no non-trivial congruence (which means, either $\overline{G}$ is transitive and primitive, or $k = 2$ and $\overline{G} = 1$); if primitive, it is necessarily basic.

4.5 The O'Nan–Scott Theorem

Before stating the theorem, we describe two more classes of groups.

- *Twisted wreath products.* These are primitive groups with non-abelian regular normal subgroups. They were omitted from the first versions of the O'Nan–Scott Theorem, but their existence was pointed out by Michael Aschbacher. They are not basic. The smallest twisted wreath product has degree $60^6 = 46\,656\,000\,000$.

- *Almost simple groups.* G is *almost simple* if $T \leq G \leq \mathrm{Aut}(T)$ for some non-abelian simple group T. (This is a specification of an abstract group; nothing is said about how it acts as a permutation group.)

We break the statement of the theorem into two parts. The first describes basic groups, and the second describes the reduction to basic groups in terms of the socle.

Theorem 4.6 *Let G be a basic primitive permutation group. Then either G is of affine or diagonal type, or G is almost simple.*

Note that for the affine and diagonal groups, the action of the socle is specified precisely; but nothing is said about the action in the almost simple case.

Theorem 4.7 *Let G be primitive but not basic, say $G \leq G_0 \,\mathrm{Wr}\, S_k$, where the wreath product has its product action, G_0 is the group induced on a fibre by its stabiliser, and G_0 is basic. Then the socle of G is N^k, where N is a normal subgroup of G_0. So in particular, either*

(a) $\mathrm{Socle}(G) = \mathrm{Socle}(G_0)^k$, *or*

(b) G_0 has two regular normal subgroups N and N', and $\mathrm{Socle}(G) = N^k$ *is regular.*

Note that N^k has its product action on Γ^k. Case (b) occurs in twisted wreath products (and only in these).

Sketch proof. Let N be the socle of the primitive group G. We have already seen that one of the following must occur (see Theorem 4.4):

(a) N is elementary abelian and regular;

(b) N is the product of two non-abelian regular normal subgroups;

(c) N is non-abelian and minimal normal and regular;

(d) N is non-abelian and minimal normal and not regular.

In case (a), G is affine. In case (b), if the minimal normal subgroups are simple, then G is of diagonal type; otherwise, it can be shown that G is contained in the wreath product of a diagonal group (with two minimal normal subgroups) with a transitive group. In case (c), it can be shown that G is of twisted wreath product type, and a careful analysis of this type shows that it is non-basic.

The most interesting case is (d). We begin with the following remark. Let $G = G_1 \times \cdots \times G_m$ be a group, and $H_i \leq G_i$ for $i = 1, \ldots, m$. Let Ω_i be the coset space $H_i \backslash G_i$, and Ω the coset space $H \backslash G$, where $H = H_1 \times \cdots \times H_m$. Then the action of G on Ω is isomorphic to the product of the actions of G_i on Ω_i for $i = 1, \ldots, m$.

Let $N = T_1 \times \cdots \times T_r$, where $T_1, \ldots, T_r$ are isomorphic non-abelian simple groups. Let $\mathcal{T} = \{T_1, \ldots, T_r\}$. We may assume that $r > 1$, since if $r = 1$ then G is almost simple.

Now G acts transitively on both Ω (in the given action) and $\mathcal{T}$ (by conjugation), and the stabiliser of an element of $\mathcal{T}$ is transitive on Ω (since N is transitive on Ω, and it fixes every element of $\mathcal{T}$). So G_α acts transitively on $\mathcal{T}$.

We claim that N_α is a maximal G_α-invariant subgroup of N. For, if $N_\alpha < K < N$ where K is G_α-invariant, then $G_\alpha < KG_\alpha < G$, contradicting the primitivity of G. Let π_i be the projection map from N to T_i, and $S_i = (N_\alpha)\pi_i$.

Case 1. $S_1 < T_1$. Then $S_i < T_i$ for all i, since G_α acts transitively on $\mathcal{T}$. Then N_α is contained in the product of its projections,

$$N_\alpha \leq S_1 \times \cdots \times S_r,$$

and the right-hand side is G_α-invariant, so we must have equality. But this means that the action of N on Ω is isomorphic to the product action of the direct product $T_1 \times \cdots \times T_r$, where T_i acts on the cosets of S_i for all i. Hence G is non-basic: in fact, $G \leq G_1 \operatorname{Wr} S_r$, where G_1 is primitive with socle T_1.

Case 2. $S_1 = T_1$. As before, this implies that $S_i = T_i$ for all i. By Exercise 4.3, this implies that there is a partition of $\mathcal{T}$ into s sets of size t, where $st = r$, such that N_α is the direct product of diagonal subgroups of the parts. Then again the action of N is determined, and we see that $G \leq G_1 \text{ Wr } S_s$, where G_1 is a diagonal subgroup of the product of t simple groups. Thus, either $s = 1$ and G is diagonal, or $s > 1$ and G is non-basic.

For a detailed proof of this theorem, see Dixon and Mortimer [64].

Using the theorem, we see that, in order to understand finite primitive permutation groups, we need to know about

- irreducible linear groups (for the affine case);
- automorphism groups of simple groups (for the diagonal case);
- maximal subgroups of almost simple groups.

Moreover, Aschbacher [6] has given an analogue of the O'Nan–Scott Theorem for linear groups. In Aschbacher's Theorem, there are more types of groups, but again the 'unknown' groups turn out to be close to simple. So our task now is to understand the finite simple groups!

4.6 Maximal subgroups of S_n

The original form of the O'Nan–Scott Theorem, as presented at the Santa Cruz conference, was a classification of the maximal subgroups of symmetric and alternating groups. We state the result for symmetric groups for tidiness: a maximal subgroup of the alternating group is the intersection of one of these groups with the alternating group.

Theorem 4.8 *A maximal subgroup of S_n is one of the following:*

(a) intransitive, $S_k \times S_l$, $k + l = n$;

(b) transitive imprimitive, $S_k \text{ Wr } S_l$ (standard action), $kl = n$;

(c) primitive non-basic, $S_k \text{ Wr } S_l$ (product action), $k^l = n$, $k \neq 2$;

(d) affine, $\mathrm{AGL}(d,p)$, $p^d = n$;

(e) diagonal, $T^k.(\mathrm{Out}(T) \times S_k)$, T non-abelian simple, $|T|^{k-1} = n$;

(f) almost simple.

In this formulation, the twisted wreath products don't arise, since they are not basic. For several applications, the above form of the theorem is the most convenient.

Not all the groups mentioned in the theorem are actually maximal. To take just a few examples:

- $S_k \times S_k$ (intransitive) $< S_k \operatorname{Wr} S_2$ (imprimitive) $< S_{2k}$;
- $S_3 \operatorname{Wr} S_l$ (product action) $< \mathrm{AGL}(l,3) < S_{3^l}$;
- $\mathrm{PSL}(2,7)$ (almost simple) $< \mathrm{AGL}(3,2) < S_8$.

All exceptions to maximality have been determined by Liebeck, Praeger and Saxl [115]. They give six tables of exceptions (one stretching over two pages), and remark that there exist arbitrarily long chains of almost simple primitive groups.

4.7 The finite simple groups

The *Classification of Finite Simple Groups*, or CFSG as we shall refer to it, is one of the most remarkable theorems ever proved. First, its length: the present version runs to an estimated 15 000 pages, spread over books, journals, computer calculations and, in at least one case, unpublished manuscripts. The proof was announced in 1980, though it was known at the time that some details (such as the proof that 'groups of Ree type' are indeed Ree groups) remained to be completed. It turned out that there was a more serious lacuna, in the treatment of 'quasi-thin groups'. It is quite impossible for a layman to judge whether a complete proof of the theorem currently exists.

The problems are being addressed in the 'revisionism' program initiated by Daniel Gorenstein and now being directed by Richard Lyons and Ronald Solomon. A self-contained proof, whose length is estimated at only 5000 pages, is being published in a number of volumes, of which the first few have already appeared (see [83]).

The impact of CFSG on many areas of mathematics is so dramatic that it cannot simply be ignored until the revised proof has appeared. Accordingly, I will use it and derive some of its striking consequences for permutation groups. Theorems proved using CFSG will always be labelled as such. I have done this for two reasons. First, simple prudence dictates this course, since we have no satisfactory proof of the theorem. Second, a result which has been proved using CFSG, but which has defied all attempts at an 'elementary' proof, probably lies quite deep; this is an interesting metamathematical observation.

Now, to the theorem:

Theorem 4.9 [CFSG] *A non-abelian finite simple group is one of the following:*

- *an alternating group A_n, $n \geq 5$;*
- *a group of Lie type;*

- *one of* 26 *sporadic groups.*

The alternating groups are familiar to us; it was known to Galois that A_n is simple for $n \geq 5$. The other classes are less familiar.

Groups of Lie type are, roughly speaking, matrix groups over finite fields; the prototype is the group $\mathrm{PSL}(n, q)$, which is the factor group of the group of $n \times n$ matrices with determinant 1 over $\mathrm{GF}(q)$ by its centre (which consists of the scalar matrices in this group). It is simple for all $n \geq 2$ and all prime powers q, except for $(n, q) = (2, 2), (2, 3)$. More generally, the groups of Lie type are conveniently divided into two types: *classical groups* (the linear, symplectic, orthogonal and unitary groups, each family parametrised by a 'rank' or 'dimension' and a field order), and the *exceptional groups*, ten families each parametrised by a field order. From many points of view, the somewhat mysterious exceptional groups behave like classical groups of fixed dimension. The names and orders of these groups are given in Table 7.1 in Chapter 7. For further details, see Carter [49], or Dieudonné [62] for the classical groups. A wealth of detail about the subgroup structure (and other properties) of the classical groups is given by Kleidman and Liebeck [110].

The *sporadic groups* are 26 specific finite simple groups, with orders ranging from 7920 to about 10^{55}. Their names and orders are given in Table 7.2. Further information is given in the **ATLAS** [54].

From CFSG and properties of the known groups, we can draw various conclusions:

Theorem 4.10 [CFSG] *(a) Any non-abelian finite simple group can be generated by two elements.*

(b) The outer automorphism group of a non-abelian finite simple group is an extension of a metacyclic group by a subgroup of S_3; in particular, it is soluble.

The fact that the outer automorphism group of a non-abelian finite simple group is soluble is known as *Schreier's Conjecture.* No direct verification of it is known.

Further properties of the simple groups will be given as needed. It is perhaps worth mentioning that, although many asymptotic consequences of CFSG would be essentially unchanged even if one (or a finite number) more sporadic groups had to be added to the list, there is currently no hope of proving that the list of sporadic groups is finite, short of proving the full CFSG as stated.

4.8 Application: Multiply-transitive groups

One of the major goals of permutation group theory since its origin has been the determination of the multiply-transitive groups or, at the very least, an absolute bound on the degree of transitivity of finite permutation groups other than symmetric and alternating groups. This goal was beyond reach until CFSG was proved, at which moment it was realised, since it follows from CFSG and previously known results.

Theorem 4.11 [CFSG] *The finite 2-transitive groups are explicitly known. In particular, the only finite 6-transitive groups are symmetric and alternating groups; and the only finite 4-transitive groups are symmetric and alternating groups and the* Mathieu groups M_{11}, M_{12}, M_{23} *and* M_{24}.

The list of 2-transitive groups appears Tables 7.3 and 7.4.

Before commenting on the proof, note that Burnside's Theorem (which is what is needed) follows immediately from the O'Nan–Scott Theorem: a 2-transitive group is basic, and no group of diagonal type can be 2-transitive.
So the 2-transitive groups are either affine or almost simple. If $G = V \rtimes G_0$ is a 2-transitive affine group, then G_0 is a linear group on V which acts transitively on the non-zero vectors of V. Such groups were determined by Huppert in the soluble case, and Hering (using CFSG) in the non-soluble case. The almost simple 2-transitive groups were determined by various people, including Maillet, Curtis, Kantor, Seitz, and Howlett. We will give an example illustrating one technique used for this in Section 4.10.
The groups with higher degrees of transitivity are now found from the list by inspection.
Some partial results can be proved by more elementary means; for example, the 3-transitive affine groups (by Cameron and Kantor) and the 3-transitive groups G in which G_α is affine (by Shult, Hering, Kantor and Seitz).
We end with one curious consequence. With one single exception (which we discussed in Section 1.16), Burnside's Theorem can be strengthened: a simple minimal normal subgroup of a 2-transitive group is itself 2-transitive.

The classification of 2-transitive groups has been one of the most useful consequences of CFSG. I mention just one application: to restricted randomisation in statistics, where the assignment of treatments to plots is randomised so that there are enough options that the statistical analysis of the result is valid, but specified 'bad patterns' are avoided. See Bailey [11].

Many related classes of primitive groups have been determined, including the rank 3 groups [108, 117, 114], the almost simple groups of odd degree [107, 116], and so on.

4.9 Application: Degrees of primitive groups

If we cannot hope to determine all the primitive groups explicitly, at least we might hope to determine their degrees. Let E be the set of positive integers n for which there exists a primitive group of degree n other than S_n and A_n. No number smaller than 5 is in E: there is just no room for primitive groups of very small degree. Mathieu discovered that every number n with $5 \le n \le 33$ lies in E. (To prove this, he was forced to discover some sporadic simple groups: the only non-trivial primitive groups of degree 22 are the Mathieu group M_{22} and its automorphism group.) It turns out that 34 is not in E, though showing this would be a daunting job without CFSG. However, it turns out that 34 is typical: the set E has density zero. More precisely, it is possible to give an asymptotic expansion for the function $e(x) = |E \cap (0, x)|$:

Theorem 4.12 [CFSG] *With the above notation,*

$$e(x) = 2\pi(x) + (1 + \sqrt{2})x^{1/2} + O(x^{1/2}/\log x),$$

where $\pi(x)$ is the number of primes less than x.

Remark. We use the standard notation of asymptotic analysis:

- $f(x) = O(g(x))$ means that $|f(x)| < cg(x)$ for some constant c;
- $f(x) = o(g(x))$ means that $|f(x)| < \epsilon g(x)$ for all $\epsilon > 0$;
- $f(x) \sim g(x)$ means that $f(x)/g(x) \to 1$.

In our case these apply as $x \to \infty$ (so that they hold for sufficiently large x).

Remark. From the Prime Number Theorem, we see that $e(x) \sim 2x/\log x$. The formula for $e(x)$ in the theorem has been given in terms of $\pi(x)$ because the difference between $\pi(x)$ and its analytic approximation $x/\log x$ swamps out the next term in the asymptotic expansion, even if the Riemann Hypothesis is true. (The Riemann Hypothesis implies that $|\pi(x) - x/\log x| = x^{1/2+o(1)}$; if it turns out to be false, then the exponent in the error term would be greater than 1/2.)

Sketch proof. First, where do the terms in the expansion come from? The cyclic group of order p has degree p, and the group $\mathrm{PSL}(2,p)$ has degree $p+1$, for each prime $p \ge 5$. The symmetric group S_m acting on 2-sets has degree the triangular number $m(m-1)/2$, and the group $S_m \mathrm{~Wr~} S_2$ in the product action has degree the square number m^2, for $m \ge 5$. The primes and primes plus one contribute $2\pi(x)$, and the square and triangular numbers contribute $x^{1/2} + (2x)^{1/2}$, to $e(x)$. The only overlap between these four series results from the fact that some numbers are both square and triangular; but this condition yields a Pell equation, whose solutions grow exponentially, so there are only $O(\log x)$ such coincidences below x. This gives a lower bound.

For the upper bound, the 'maximal subgroups' form of the O'Nan–Scott Theorem (Theorem 4.8) is suitable, since any primitive group is contained in a maximal one. What does the theorem tell us about the degrees of such groups?

- If G is not basic, then its degree is a proper power. The number of proper powers below x is

$$\lfloor x^{1/2} \rfloor + \lfloor x^{1/3} \rfloor + \lfloor x^{1/4} \rfloor + \cdots = x^{1/2} + O(x^{1/3} \log x),$$

  the last part because there are at most $\log_2 x$ terms in the sum. (If a proper mth power lies below x then certainly $2^m < x$.) If higher accuracy were required, we could write $x^{1/2} + x^{1/3} + O(x^{1/4} \log x)$.

- If G is affine, then its degree is a prime power, and hence either a prime (this gives a contribution $\pi(x)$ to $e(x)$) or a proper power (this has already been counted).

- If G is diagonal, then its degree is $|T|^{k-1}$, where T is a non-abelian finite simple group. If $k = 2$, this is $|T|$. From CFSG, it is not difficult to estimate the contribution to $e(x)$. Indeed, as a glance at the tables in Dickson [61] or the **ATLAS** [54] strongly suggests, 'most' finite simple groups are $\mathrm{PSL}(2,p)$ for some prime p; these have order $(p^3 - p)/2$, and so the contribution is $cx^{1/3}/\log x$. If $k > 2$, the degree is a proper power, and these have already been counted.

- Suppose that G is almost simple. This is the most difficult case; we have to work through the known families of simple groups and estimate the contribution of each to $e(x)$. It turns out that alternating groups contribute $(2x)^{1/2} + O(x^{1/3})$, the leading term arising from S_m or A_m on 2-sets; groups of Lie type contribute $\pi(x) + O(x^{1/2}/\log x)$, the leading term arising from $\mathrm{PSL}(2,p)$ or $\mathrm{PGL}(2,p)$, in their actions on the projective line (with degree $p+1$); and the sporadic groups contribute only a constant.

For further details, see the paper by Cameron, Neumann and Teague [45].

4.10 Application: Orders of primitive groups

The problem of the orders of subgroups of symmetric groups has a long history. The Grand Prix question of the Académie des Sciences, Paris, in 1860 asked:

How many distinct values can a function of n variables take?

In other words, what are the possible indices of subgroups of S_n? (If f is a function of n variables, and G is the group of permutations of the arguments of f which leave the value unchanged, then the distinct values of f correspond to cosets of G in S_n.) Jordan, Mathieu and Kirkman submitted entries for the prize, but it was not awarded. Indeed, in the form stated, the problem is well beyond present-day resources.

I am grateful to Joachim Neubüser for the historical comments which follow.

A really amusing description of the history of the problem is given in Appendix A, pp. 122–124 of Short [166]. It is described as a history of errors, corrections, and corrections of corrections, involving almost all of the heroes of the nineteenth century.

The first good list of primitive groups was by Charles Sims leading up to degree 50. Short then determined (generators of) the soluble primitive groups up to degree 243. Dixon and Mortimer [63] described the non-affine primitive groups. Generators for these were determined (and a very few minor errors corrected) in the PhD thesis of Theissen [173]. He also determined the non-soluble affine primitive groups of degree up to 243, so that the primitive groups are now completely known up to that degree. Generators for these groups are included in GAP and MAGMA.

Hulpke [95] found the transitive groups up to degree 31, and all permutation groups (all subgroups of S_n, up to conjugacy) up to degree 10. Details can be found on his World Wide Web page at the URL

`http://www-groups.dcs.st-and.ac.uk/~ahulpke/smalldeg.html`

which includes, among other things, a 'transitive group of the day'.

Let us make some simplifications. First, we will treat only primitive groups, and ignore the reductions from arbitrary groups. Second, we will simply ask for good upper bounds for the orders of primitive groups other than S_n and A_n. With these restrictions, the state of current knowledge is satisfactory, as we will see.

During the 'classical' period, a number of bounds were proved, culminating in exponential bounds c^n by Wielandt [187] (for groups which are not 2-transitive) and Praeger and Saxl [149] (in general): if G is primitive and does not contain the alternating group, then $|G| \leq 4^n$.

Babai [7] introduced novel combinatorial methods into the problem. By using techniques of probabilistic combinatorics, he was able to bound the size of a base in a primitive but not 2-transitive group G, and hence to show that such a group satisfies $|G| \leq n^{4n^{1/2} \log n}$. This is essentially best possible. For let G be the symmetric group S_m in its action on 2-sets. Then $n = m(m-1)/2$, and $|G| = m! > (m/e)^m$; so $|G| > n^{cn^{1/2}}$. A similar assertion holds for the group S_m Wr S_2 (in the product action).

Subsequently, Pyber [151] used similar methods to prove an even stronger bound for 2-transitive groups G other than S_n and A_n: he finds that $|G| \leq n^{c(\log n)^2}$, which is again best possible up to a factor of $\log n$ in the exponent. (Consider the affine groups $\mathrm{AGL}(d, 2)$, with $n = 2^d$: the order is about $cn^{1+\log_2 n}$.)

Given that these bounds are essentially best possible, what more needs to be said? Using CFSG, one can do significantly better, by allowing a list of 'known' exceptions. The result is limited only by the amount of vagueness allowed in the list of exceptions. As we will see, there is a 'bootstrap principle', by means of which very poor bounds for the orders of subgroups of S_m can be converted into much stronger bounds on orders of subgroups of S_n for n much greater than m. I will illustrate these points with a sketch proof of the following theorem taken from [33].

Theorem 4.13 [CFSG] *There is a constant c such that, if G is primitive of degree n, then either*

(a) $A_m^l \leq G \leq G_0$ Wr S_l in the product action, where G_0 is S_m or A_m on k-sets; or

(b) $|G| \leq n^{c \log n}$.

Remark. For fixed k and l and large m, the groups in (a) have order very roughly $n^{n^{1/kl}}$. These include the symmetric and alternating groups themselves, as well as the 'large' groups S_m on 2-sets or S_m Wr S_2 that we have already encountered. In computational permutation group theory, it is customary now to divide primitive groups into two classes as in the theorem: the first class are known well enough that they can be dealt with; the second are small enough that the standard algorithms handle them efficiently. See [10], or several papers in the DIMACS *Groups and Computation* proceedings [75], for a survey of fast algorithms for dealing with small groups. (These groups are called 'small base groups', since they have bases of small size. See Section 4.13 below for the relationship between base size and order, and the significance of base size for computation.)

Sketch proof. We begin with an observation: if $n = c^m$, then $m = \log n / \log c$, and so

$$m! \leq m^m \leq n^{c' \log\log n}.$$

Now suppose that G is non-basic, say $G \leq G_0$ Wr S_l, with degree $n = n_0^l$. Then

$$|G| \leq |G_0|^l l! \leq |G_0|^l n^{c' \log\log n}.$$

Let us suppose that we can prove that, for basic groups G_0, either G_0 is S_m or A_m on k-sets, or $|G_0| \leq n^{c'' \log n_0}$. In the first case we have conclusion (a) of the theorem, while the second gives

$$|G| \leq (n_0^{c'' \log n_0})^l n^{c' \log \log n} = n^{(c''/l) \log n + c' \log \log n} \leq n^{c \log n}.$$

So we have to prove the result for basic groups.

If G is affine, then $n = p^d$ and $|G| \leq p^{d(d+1)} = n^{1+\log_p n}$.

If G is diagonal, then $n = |T|^{k-1}$ and $|G| \leq |T|^k |\operatorname{Out}(T)| k!$. Now it follows from CFSG that the outer automorphism groups of finite simple groups are very small. The bound $|\operatorname{Out}(T)| < |T|$ is extravagantly bad, but easily good enough for our needs. Using the fact that $k+1 \leq 3(k-1)$ for $k \geq 2$, and our earlier observation, we see that $|G| \leq n^{3+c' \log \log n}$.

Suppose that G is almost simple. If the socle of G is sporadic, then certainly the bound holds for some c''. If the socle is classical, then the order of G is bounded by q^{ar^2}, where q is the field of definition, r the Lie rank, and a a constant. A result of Cooperstein [55] gives the smallest degree of a permutation representation of G; this is roughly q^{br} for some constant b, and is a lower bound for n. So we obtain $|G| \leq n^{c' \log n}$ again. A similar but cruder bound is easily derived for the exceptional groups.

We are left with the case that the socle of G is A_m, that is, that G is A_m or S_m (excluding the case $m = 6$, the unique degree where $\operatorname{Aut}(A_m) \neq S_m$: this case can be treated as sporadic). Let $H = G_\alpha$, a maximal subgroup of S_m or A_m. Now we use the 'maximal subgroup' form of the O'Nan–Scott Theorem (Theorem 4.8). If H is intransitive, it is the stabiliser of a k-set, and G is S_m or A_m on k-sets. If H is transitive and imprimitive, or if it is primitive, then its index in S_m or A_m (which is n) is very large, and the order of S_m is small as a function of m.

It is at this step that the bootstrap principle comes into play. If H is imprimitive, so that $H = S_a \operatorname{Wr} S_b$ with $ab = m$, then we know everything explicitly. If H is primitive, only very weak bounds on its order suffice. For example, an ancient bound due to Bochert (see Wielandt [186], or Exercise 4.22) asserts that $n = |S_m : H| \geq \lceil m/2 \rceil!$, so that $|S_m| = m! < n^3$. We have converted a superexponential bound on the order of a subgroup of S_m to a polynomial bound on the order of a certain kind of subgroup of S_n!

As noted earlier, much stronger bounds can be proved if we are prepared to allow longer lists of exceptions. As we saw, the 'top group' of a wreath product, or the S_k factor in a diagonal group, only contributes $n^{c \log \log n}$ to the order. Affine groups, and classical groups in 'known' actions, have order $n^{c \log n}$, so these must be placed in the list of exceptions. Similarly for S_m or A_m in the case where H is transitive but imprimitive in the earlier analysis. With these exceptions we can obtain a bound of the form $n^{c \log \log n}$.

Improving this in general would require bounding the orders of transitive groups (since the top group in the wreath product is only required to be transitive). But we can do better for almost simple groups (see Liebeck [113]):

Theorem 4.14 *There is a constant c such that, if G is an almost simple primitive group of degree n, then one of the following holds:*

(a) G is S_m or A_m, acting on k-sets or on partitions into l sets of size k, for some k, l;

(b) G is a classical group, acting on an orbit of subspaces (or pairs of subspaces of complementary dimension) in its natural module;

(c) $|G| \leq n^c$.

In fact, the worst case in this theorem is the Mathieu group M_{24}, with $c = 6.077\,948\,094$ (Martin Liebeck, unpublished). Perhaps it is true that $|G| \leq n^5$ holds if we permit finitely many more exceptions. (The action of $E_8(q)$ on the cosets of a maximal parabolic subgroup may be extremal here; its order is roughly $n^{4.28}$. I am grateful to Gary Seitz and a referee for this observation.)

We now illustrate how even very weak bounds for orders of primitive groups can be used in determining the 2-transitive groups. We will show that the symmetric group S_n has a 2-transitive action not isomorphic to the natural action only for $n \leq 6$.

Theorem 4.15 *Let G be 2-transitive on Ω and have a non-identity conjugacy class C. Then $|\Omega| \leq 1 + |C|$.*

Proof. Suppose that some element $g \in C$ maps α to β. Take any point $\gamma \neq \alpha$, and let x fix α and map β to γ. Then $x^{-1}gx \in C$ maps α to γ. So the number of points different from α does not exceed $|C|$.

Now suppose that S_n acts 2-transitively on Ω. Using the conjugacy class of transpositions in S_n shows that $|\Omega| \leq (n^2-n+2)/2$. Moreover, the stabiliser of a point of Ω is a maximal subgroup G of S_n. If it is intransitive or imprimitive, then Ω is isomorphic to a set of subsets or partitions of $\{1, \ldots, n\}$; these cases are easily dealt with. If G is primitive, then its order cannot be as large as $2n!/(n^2 - n + 2)$ except for very small values of n.

For groups of Lie type, there are also fairly small conjugacy classes, consisting of 'long root elements' (these are the transvections in the case of the groups $\mathrm{PSL}(n, q)$), to which a similar argument can be applied.

4.11 Application: The length of S_n

Recall that we defined the *length* of a finite group G, denoted by $l(G)$, to be the length of the longest chain of subgroups in G. Theorem 1.14 gives the formula for the length of S_n:

$$l(S_n) = \lceil 3n/2 \rceil - b(n) - 1,$$

where $b(n)$ is the number of ones in the base 2 representation of n. Let $f(n)$ denote this function. We showed that there is a chain of length $f(n)$, and deferred the proof that no longer chain exists.

We prove this by induction, assuming that $l(S_m) = f(m)$ for all $m < n$. Let

$$S_n > G > \cdots > 1$$

be a longest chain of subgroups in S_n, with length N. Then G is a maximal subgroup of S_n, and so is of one of the six types in Theorem 4.8. We consider these in turn. In each case we derive a bound for N, and some calculation shows that it implies $N \leq f(n)$.

- G intransitive, $G = S_k \times S_l$, with $k + l = n$. Then $N = f(k) + f(l) + 1$.
- G imprimitive, $G = S_k \operatorname{Wr} S_l$, with $kl = n$. Then $N = lf(k) + f(l) + 1$.
- G non-basic, $G = S_k \operatorname{Wr} S_l$, with $k^l = n$. Here N is the same as in the previous case, but n is larger; so no calculation is required!
- G is affine. Then $|G| \leq n^{1+\log_p n}$; and clearly $N = 1 + l(G) \leq 1 + \log_2 |G|$.
- G is diagonal, $G = T^k.(\operatorname{Out}(T) \times S_k)$. Then $N = 1 + l(G) \leq 1 + k\log_2 |T| + \log_2 |\operatorname{Out}(T)| + f(k)$.
- G is almost simple, $G \neq A_n$. Use $N \leq 1 + \log_2 |G|$ and the bounds of the preceding section.
- $G = A_n$. Now let the series continue

  $$S_n > A_n > H > \cdots > 1,$$

  and apply the same analysis to H.

The point is that, if we use the 'elementary' bounds of Babai and Pyber, then the theorem can still be proved in principle, since

$$\log_2(n^{4n^{1/2}\log n}) = o(n);$$

but we have to analyse many more small cases. (A rough estimate suggests that it would be necessary to deal with primitive groups of degree up to about 10^6, whereas using CFSG we only need to go up to degree 30 or so.)

4.12 Application: Distance-transitive graphs

In the preceding chapter, we set up the situation roughly as it existed in the theory of distance-transitive graphs prior to the announcement of CFSG in 1980. Now we can follow the story forward to the present.

The *Sims Conjecture* was proved in the early 1980s [46]:

Theorem 4.16 [CFSG] *There is a function f with the property that, if G is a finite primitive permutation group in which G_α has an orbit of length k (other than $\{\alpha\}$), then $|G_\alpha| \leq f(k)$.*

There is not room to give the proof here, but the main lines can be indicated. The first, essential ingredient is a theorem of Thompson, refined by Wielandt, which asserts that there is a function g such that, with the hypotheses of the theorem, G_α has a normal subgroup of index at most $g(k)$ whose order is a power of a prime p. Also, using the O'Nan–Scott Theorem, it is enough to prove the conjecture in the case where G is almost simple. So, if $|G_\alpha| > g(k)$, then G is an almost simple group and $H = G_\alpha$ is a *local subgroup* of G (the normaliser of a p-subgroup) which is maximal in G. Moreover, $|H : H \cap g^{-1}Hg| = k$ for some $g \in G$. So the argument requires detailed knowledge of how local subgroups of almost simple groups intersect their conjugates.

The truth of the Sims conjecture makes it feasible to complete Smith's program:

Theorem 4.17 [CFSG] *For given $k > 2$, there are only finitely many finite distance-transitive graphs of valency k.*

A feature of the proof was the use of infinite techniques. The argument depends on a theorem of Macpherson [126], which completely determines the infinite distance-transitive graphs of finite valency. These graphs have a particularly simple structure. For each pair s, t of positive integers greater than 1, there is a graph $M(s,t)$ built from complete graphs of size s, joined together so that t of them pass through any vertex, and the cliques are arranged in treelike fashion so that there are no cycles in the graph except for those contained in a clique. The number of vertices at distance d from a given vertex is $t(t-1)^{d-1}(s-1)^d$; this grows exponentially with d unless $s = t = 2$ (in which case the graph is an infinite path). Figure 4.4 shows part of Macpherson's graph $M(4,2)$.

Now suppose that there are infinitely many distance-transitive graphs of given valency $k > 2$. By Smith's reduction, there will be infinitely many primitive distance-transitive graphs of the same valency; so we may assume that all our graphs are primitive. By the truth of the Sims conjecture, the

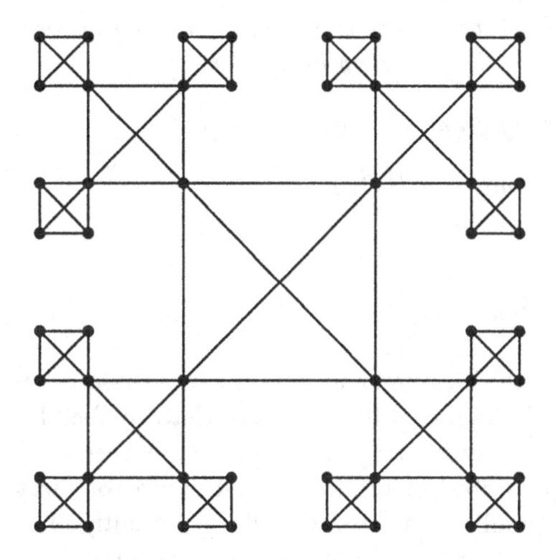

Figure 4.4: A distance-transitive graph

number of vertices at distance d from a given one is bounded by a function of k, independent of d. Choose d large enough that this bound is smaller than the number of vertices at distance d from a given one in any Macpherson graph $M(s,t)$ of valency k. Then there is a number l such that infinitely many finite examples of the following situation exist:

> Γ is a primitive distance-transitive graph with valency k and having l vertices at distance d from any given vertex.

Now we reach a contradiction as follows. The displayed conditions (apart from primitivity) can be expressed in the first-order language of groups acting on graphs. (See Section 5.3 for a brief sketch of first-order logic.) By the Compactness Theorem (Theorem 5.5), since infinitely many finite graphs satisfy these conditions, there is an infinite graph which satisfies them. But, by our choice of d, this contradicts Macpherson's Theorem.

Attention now turns to the classification of distance-transitive graphs. This is not yet complete. However, the Sims conjecture turned out to be a red herring: progress has come from looking at the whole group, not the point stabiliser.

The problem breaks into two parts, as suggested by Smith's Theorem 3.19: first, determine the primitive distance-transitive graphs; then, investigate the bipartite doubles and antipodal covers of the known graphs. The first part of the program is almost complete. The analysis using the O'Nan–Scott Theorem was given by Praeger, Saxl and Yokoyama [150]:

Theorem 4.18 *Let Γ be a finite primitive distance-transitive graph, and $G = \mathrm{Aut}(\Gamma)$. Then one of the following holds:*

(a) Γ is a Hamming graph $H(n,q)$;

(b) Γ is the complement of $H(2,q)$;

(c) G is affine;

(d) G is almost simple.

So we are left with the affine and almost simple cases. Two recent surveys are Ivanov [99] and Praeger [148]. We note that, in the almost simple case, groups whose socle is alternating or sporadic have been dealt with [98, 100], and substantial progress has been made on groups of Lie type. There are also many partial results on bipartite doubles and antipodal covers of known graphs, for which we also refer to the surveys [99, 148].

One technique used for this is simple enough to outline here. Let G be an almost simple group, r the number of conjugacy classes of G, and $f_1, \ldots, f_r$ the degrees of the irreducible characters. As we saw in the last chapter, if G acts distance-transitively on a graph, then the permutation character of G in the vertex set of the graph is multiplicity-free, so its degree is not greater than $M = \sum_{i=1}^r f_i$. Now $\sum_{i=1}^r f_i^2 = |G|$; so Cauchy's inequality gives $M \leq \sqrt{r|G|}$. So the vertex stabiliser has order at least $\sqrt{|G|/r}$. Since r is much smaller than $|G|$, the vertex stabiliser is a very large subgroup. The possibilities can be listed, by methods like those we have seen for the symmetric group. Then we have to check some known permutation actions of G to see whether the permutation character is multiplicity-free, and if so, whether one or more of the orbital graphs is distance-transitive.

4.13 Bases

Recall that a *base* for a permutation group G is a sequence $(\alpha_1, \alpha_2, \ldots, \alpha_d)$ whose pointwise stabiliser in G is the identity. A base is *irredundant* if no base point is fixed by the stabiliser of the earlier points in the base. (Such 'redundant' points could be dropped and a smaller base obtained.)

There are practical reasons for trying to find small bases. In computing with a permutation group, any element is uniquely specified by its action on a base. (For if $\alpha_i g = \alpha_i h$ for all i, then gh^{-1} fixes every base point, and so is the identity; so $g = h$.) So the smaller the base, the less memory is needed to store group elements.

If G has a base of length d, then $|G| \leq n^d$, since there are at most n^d different d-tuples of points of Ω. (In fact, Babai's bound $|G| \leq n^{4n^{1/2}\log n}$ for the order of a uniprimitive permutation group was proved by showing that such a group has a base of size at most $4n^{1/2}\log n$.) On the other hand, if G has an irredundant base of length d, then $|G| \geq 2^d$: for, if $G(i)$ denotes the stabiliser of the first i basepoints, then $G(i) < G(i-1)$ for all i, whence $|G(i-1) : G(i)| \geq 2$, and $|G| \geq 2^d$. Hence we obtain:

Theorem 4.19 *Let G be a permutation group with minimal base size b. Then any irredundant base for G has size at most $b\log_2 n$.*

Proof. We have $|G| \leq n^b$, and the size of an irredundant base is at most $\log_2 |G|$.

A small modification lets us do much better. The *greedy algorithm*, due to Kenneth Blaha [20], says: always choose the basepoint α_i from a longest orbit of $G(i-1)$ (the stabiliser of the preceding basepoints). The heuristic explanation is that we want to reach the identity as quickly as possible by going down a chain of point stabilisers, so we move to the smallest possible subgroup (that is, largest possible index in its predecessor) at each step. We say 'a longest orbit' since, at a given stage, there may be several orbits all of the same size.

Theorem 4.20 *Let G be a permutation group with minimal base size b. Then any base found by the greedy algorithm has size at most $b(\log\log n + c)$, where c is a constant.*

Proof. First, we observe:

For any subgroup H of G, H has an orbit of size at least $|H|^{1/b}$.

For H has an orbit of length $|H|$ on b-tuples, namely, the images of a base; this would not be possible if every H-orbit was smaller than $|H|^{1/b}$. This means that, if α is chosen from the largest orbit of H, then $|H_\alpha| \leq |H|^{1-1/b}$.

Now use the greedy algorithm to choose $b\log\log n$ base points. Their stabiliser has order at most

$$|G|^{(1-1/b)^{b\log\log n}} \leq (n^b)^{\mathrm{e}^{-1/b\cdot b\log\log n}} = n^{b/\log n} = \mathrm{e}^b;$$

so the choice of any $b\log_2 \mathrm{e}$ further basepoints by any irredundant method completes a base.

However, for primitive groups, more is conjectured to be true:

Conjecture. There is an absolute constant c such that, if a primitive group has minimal base size b, then any base produced by the greedy algorithm has size at most cb.

Moreover, the constant c should be $9/8$. The extremal cases should be our old friends, S_m acting on 2-sets.

For suppose that we have fixed a certain number of 2-sets covering a subset S of size k of the original set Ω. A 2-set contained in S lies in an orbit of length at most 4; one meeting S in one point lies in an orbit of length $n-k$ or $2(n-k)$; and one disjoint from S lies in an orbit of length $(n-k)(n-k-1)/2$. So the greedy algorithm chooses disjoint pairs until at most five points remain uncovered. At some stage it must return and choose further pairs to fix the points in these disjoint pairs. Thus, apart from a bounded part of Ω, three base pairs cover four points of Ω, and the base has size $3m/4+c$ (see Figure 4.5).

Figure 4.5: A greedy base

However, it is possible to use two pairs to fix three points of Ω, and obtain a base of size $2m/3+c$ (see Figure 4.6): this is the smallest base.

Figure 4.6: A minimal base

Recall Theorem 4.14, where with known exceptions we saw that the order of an almost simple group is bounded by n^c for some absolute constant c. This might suggest that, with the same exceptions, an almost simple primitive group has base size bounded by an absolute constant. Probably this constant is 7, and the extreme case is the Mathieu group M_{24}. But perhaps, with finitely many exceptions, the correct bound is actually 5. The existence of such a constant has very recently been proved by Martin Liebeck and Aner Shalev [119]; but their proof, which is probabilistic, does not give an explicit bound for the constant.

On the other hand, Pyber [152] has shown the following result.

Theorem 4.21 *There exists a constant β such that almost all subgroups of S_n (a proportion tending to 1 as $n \to \infty$) have minimal base size at least βn.*

Note that this theorem does not use CFSG: the 'elementary' estimates for orders of primitive groups given by Babai and Pyber suffice. Pyber's paper contains many further results and problems on 'asymptotic permutation group theory', and is well worth reading.

4.14 Geometric groups and IBIS groups

Bases for a group possess some of the features of bases in a vector space. (Indeed, a basis for a vector space V is an irredundant base for the general linear group $\mathrm{GL}(V)$, and conversely.) However, some familiar properties of bases in vector spaces do not extend to bases for permutation groups. There may be bases of different cardinality, as we have seen. Re-ordering an irredundant base always gives a base, but it may no longer be irredundant. And not every bijection between bases extends to a group element.

The most straightforward connection between a permutation group G and its bases occurs when G permutes its irredundant bases transitively. Such a group is called a *geometric group*. This term is not entirely satisfactory, since there are many different ways in which a group may be said to be 'geometric'. However, geometric groups do indeed preserve a geometry, as we shall see later. We define the *rank* of a geometric group to be the cardinality of a basis. (Be careful not to confuse this with the rank of an arbitrary permutation group.)

A sharply t-transitive group is geometric: its bases are all the t-tuples of distinct elements, and its rank is t. As we noted at the start of the section, the general linear group $\mathrm{GL}(V)$ is geometric; its rank is the dimension of V, and its irredundant bases are the bases of V.

There is no hope of determining all the geometric groups, since any regular group is geometric. However, using CFSG, Maund [135] has determined all finite geometric groups of rank greater than 1. The classification for sufficiently large rank (at least 7) was given by Zil'ber, not using CFSG, for applications in model theory.

Infinite geometric groups can also be defined. We say that such a group has *finite type* if the number of fixed points of any non-identity element is finite. Now the following result of Cameron [35] generalises Tits' Theorem 1.11 on the non-existence of sharply 4-transitive groups. The proof follows closely that of Tits.

Theorem 4.22 *There is no infinite geometric group of finite rank at least* 4 *and finite type.*

The geometry associated with a geometric group is a matroid. See the Appendix to this chapter for more discussion of matroids. For the present, a matroid on a set Ω is specified by a collection of subsets called *bases*, no basis containing another, such that the *Exchange Axiom* holds: if B_1 and B_2 are bases and $x \in B_1 \setminus B_2$, then there exists $y \in B_2 \setminus B_1$ such that $B_1 \setminus \{x\} \cup \{y\}$ is a basis.

A generalisation of the notion of a geometric group was given by Cameron and Fon-Der-Flaass [41], who proved the following theorem.

Theorem 4.23 *The following conditions on a finite permutation group are equivalent:*

(a) all irredundant bases have the same size;

(b) the irredundant bases are invariant under re-ordering;

(c) the irredundant bases are the bases of a matroid.

Proof. (a) implies (b): Suppose that all irredundant bases have the same size. If an irredundant base is re-ordered, the result is a base, and can be made irredundant by deleting redundant elements (fixed by the stabiliser of their predecessors). But no deletions can be made, or a smaller irredundant base would be found.

(b) implies (c): Suppose that (b) holds, so that we can regard irredundant bases as unordered sets. Axiom (M1) is clear. Suppose that B_1 and B_2 are bases and $x \in B_1 \setminus B_2$. By re-ordering, we can suppose that x is the last element of B_1. Now consider the sequence formed by the elements of $B_1 \setminus \{x\}$ followed by those of B_2. This is a base, and can be made irredundant by deletion of some elements, none of which will come from $B_1 \setminus \{x\}$. But the pointwise stabiliser H of $B_1 \setminus \{x\}$ must act semiregularly on its non-fixed points. (For if $1 < H_z < H$ for some point z, then (B_1, z, x) is an irredundant base, but (B_1, x, z) is not.) So only one point y from $B_2 \setminus B_1$ is required to complete an irredundant base.

(c) implies (a): This holds since (a) is a property of matroids.

A permutation group G satisfying the equivalent conditions (a)–(c) of Theorem 4.23 is called an *IBIS group*. (This is an acronym for *Irredundant Bases of Invariant Size.*) As for a geometric group, we define the *rank* of an IBIS group to be the cardinality of a base.

Problem. Classify the IBIS groups.

This is likely to be considerably harder than finding the geometric groups. Any semiregular group is an IBIS group of rank 1. But the direct product of IBIS groups is an IBIS group, whose rank is the sum of the ranks of the factors. So we cannot even hope for a complete classification of IBIS groups of sufficiently large rank! But it might be reasonable to pose this question for transitive IBIS groups (or perhaps primitive ones).

There are several related open problems in this area; very little is known.

Problem. Which groups have the property that all bases chosen by the greedy algorithm have the same size? What combinatorial structure is formed by the 'greedy bases' in such groups? Which groups permute their greedy bases transitively?

Problem. Investigate variants of the greedy algorithm, such as the following version due to Tracey Maund: *Always choose the next basepoint so as to maximise the number of orbits of its stabiliser.*

Problem. A geometric group is an automorphism group of a matroid which permutes the ordered bases regularly. The converse is false: determine the larger class of groups with this property. The still larger class of groups permuting the ordered bases transitively cannot reasonably be classified, but the matroids can be determined (since the group acts 2-transitively on the set of rank 1 flats). However, the question of which groups permute *unordered* bases transitively seems considerably harder.

4.15 Appendix: Matroids

There are many possible definitions of a matroid, in terms of the bases, the independent sets, the rank function, the circuits (minimal dependent sets), the hyperplanes, the subspaces, etc. See Welsh [181] or Oxley [145] for this. We give a definition in terms of bases.

A *matroid* is a pair $(X, \mathcal{B})$, where X is a set and $\mathcal{B}$ a family of subsets of X having the following properties.

(M1) No member of $\mathcal{B}$ contains another.

(M2) (The *Exchange Axiom*) If $B_1, B_2 \in \mathcal{B}$ and $x \in B_1 \setminus B_2$, then there exists $y \in B_2 \setminus B_1$ such that $B_1 \setminus \{x\} \cup \{y\} \in \mathcal{B}$.

The definition, due to Whitney, represents an axiomatisation of the notion of basis in a vector space. Many of the usual properties of vector space bases can be developed in this context: for example, any two bases have the same number of elements (since any basis can be converted into any other by a sequence of exchanges as in (M2)). Now it is easy to show that the bases of a geometric group form a matroid.

Another term used for matroids is 'combinatorial geometries'. Michel Deza defined the concept of a 'permutation geometry', in which a set and its subsets are replaced by a permutation and its 'partial permutations'. Sets of permutations which give rise to such geometries were called 'geometric sets' of permutations; if they happened to be groups, they were 'geometric groups'. Deza's definition turned out to be equivalent to the one given here.

We defined a matroid in terms of its bases. In a matroid, we make the following definitions:

- A set is *independent* if it is contained in a basis.
- The *rank* of a set is the size of the largest independent set it contains.
- A subset S is a *subspace*, or *flat*, if the addition to S of any point outside it increases the rank.

Now it turns out that any of these concepts can be used to axiomatise a matroid. Different viewpoints are useful in different contexts. For our purposes, the most important, apart from bases as described above, involves the subspaces. Let $\mathcal{F}$ be a family of subsets of a set X. Assume that X is a member of $\mathcal{F}$. We say that

- $\mathcal{F}$ is a *meet-semilattice* if the intersection of any subfamily of $\mathcal{F}$ belongs to $\mathcal{F}$;
- $\mathcal{F}$ is *ranked* (and its *rank* is r) if every maximal chain of members of $\mathcal{F}$ contains exactly $r+1$ elements.

Now, if $\mathcal{F}$ is a ranked meet-semilattice, then for every subset Y of X, there is a unique minimal set $\langle Y\rangle \in \mathcal{F}$ containing Y; it is the intersection of all the members of $\mathcal{F}$ which contain Y. The *rank* $\rho(Y)$ of Y is one less than the number of elements in any longest chain of elements of $\mathcal{F}$ with greatest element $\langle Y\rangle$.

Theorem 4.24 *A family $\mathcal{F}$ of subsets of X consists of the flats of a matroid if and only if it is a ranked meet-semilattice with the property that, if $F \in \mathcal{F}$ and $x \notin F$, then $\rho(F \cup \{x\}) = \rho(F)+1$; in other words, the sets $F' \setminus F$, where $F' \in \mathcal{F}$, $F' \supseteq F$, and $\rho(F') = \rho(F)+1$, partition $X \setminus F$.*

Informally, this means that the geometry looks as follows, where we call flats of rank $1, 2, 3, \ldots$ *points*, *lines*, *planes*, ... : two points lie in a unique line; a line and a point not on it lie in a unique plane; and so on.

One property of vector space bases which is not shared by arbitrary matroids is that any bijection between bases extends to an automorphism of the matroids. Of course, this property holds for the matroids arising from geometric groups.

Some features of the matroid are easily seen. For example, the subspaces (flats) of the matroid are the fixed point sets of elements (or subgroups) of the group.

A matroid can also be characterised locally. The geometry of flats is a special kind of *Buekenhout geometry*, whose diagram is shown in Figure 4.7.

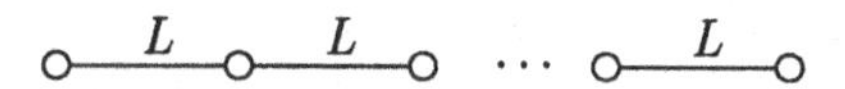

Figure 4.7: Diagram for matroids

The nodes in this diagram stand for the different types of geometric objects: in our case, the flats of rank $0, 1, \ldots, r-1$. The stroke labelled L stands for the class of *linear spaces*, or geometries of points and lines in which any two points lie on a unique line. The diagram carries two pieces of information about the incidence of flats, where two flats are *incident* if one contains the other:

- if $i < j < k$, and F_i, F_j, F_k are flats of rank i, j, k such that F_i and F_k are both incident with F_j, then F_i and F_k are incident with one another;

- if F and F' are flats of rank i and $i+3$ which are incident, then the flats of rank $i+1$ and $i+2$ incident with F and F', with the induced incidence relation, form the points and lines of a linear space.

For more information about matroids, we refer to the books of Welsh [181] or Oxley [145], or to the 'matroid page' on the World Wide Web, at

`http://members.aol.com/matroids/index.htm`

4.16 Exercises

4.1. Prove Jordan's Theorem for infinite permutation groups.

4.2. Prove by induction on n that the alternating group A_n is simple for all $n \geq 5$. [*Hint:* The induction begins with Wagner's Theorem. Now suppose that A_n is simple and let N be a non-trivial normal subgroup of A_{n+1}. Then N is transitive, and N_α is a normal subgroup of A_n. If $N_\alpha = 1$, then N is regular; show that this is impossible. So, by induction, $N_\alpha = A_n$, and $N = A_{n+1}$.]

4.3. Let G_i be a non-abelian finite simple group for $i = 1, \ldots, n$. Let H be a subdirect product of $G_1, \ldots, G_n$ (a subgroup of the direct product which projects onto each factor). Show that there is a partition of $\{1,\ldots,n\}$ into subsets $P_1,\ldots,P_k$ such that

(a) for $1 \le h \le k$ and $i_1, i_2 \in P_h$, the groups G_{i_1} and G_{i_2} are isomorphic;

(b) $H = D_1 \times \cdots \times D_k$ where, for $1 \le h \le k$, D_h is a diagonal subgroup of $\prod_{i \in P_h} G_i$.

4.4. Prove that, for any given s, there are only finitely many groups G with s conjugacy classes.

4.5. Let T be a non-abelian finite simple group, and let s be the number of orbits on $T \setminus \{1\}$ of the group $\langle \mathrm{Aut}(T), x \mapsto x^{-1} \rangle$. Prove using CFSG that there are only finitely many possible T for each given value of s. (You need some information about the finite simple groups: see Chapter 7.)

4.6. Let G be a finite primitive permutation group, having degree n, rank r, and socle $T_1 \times \cdots \times T_m$, where $T_1,\ldots,T_m$ are isomorphic non-abelian finite simple groups. Prove that

(a) $m \le r - 1$;

(b) [CFSG] if $G \le G_0 \mathrm{Wr}\, S_l$, where G_0 is a basic group of diagonal type, then $|T_i|$ (and hence n) is bounded by a function of r.

[*Hint:* Use the result of the previous exercise, or see [33].]

4.7. If H is transitive on Γ and has rank r, show that the rank of $H \mathrm{Wr}\, S_n$ (in the product action) is $\binom{r+n-1}{n}$.

4.8. Calculate the rank of the maximal diagonal group $(A_5)^3.(C_2 \times S_3)$.

4.9. Prove that the only intransitive group which preserves no non-trivial congruence is the identity group of degree 2.

4.10. Investigate when the subgroups of S_n listed in Theorem 4.8 (a)–(e) contain odd permutations.

4.11. Prove that, if G has a non-trivial conjugacy class C and acts transitively on Ω, then G_α has an orbit different from $\{\alpha\}$ of size at most $|C|$. Use Higman's Theorem to deduce that, if G is primitive and has rank r, then $|\Omega|$ is bounded by a function of $|C|$ and r.

4.12. Find

(a) all 2-transitive actions of alternating groups;

(b) all rank 3 actions of symmetric groups.

4.13. A *transvection* is an element of $\mathrm{PGL}(n,q)$ induced by a linear map of the form $x \mapsto x + (x\phi)a$ of the vector space V, where $a \in V$, $\phi \in V'$ (the dual space of V) and $a\phi = 0$. Prove that the transvections form a conjugacy class in $\mathrm{PSL}(n,q)$ of size $(q^n - 1)(q^{n-1} - 1)/(q-1)$ if $n > 2$.

4.14. Show that the Mathieu group M_{24}, in its usual action on 24 points, is an IBIS group. What is the base size? Describe the *hyperplanes* of the corresponding matroid (the maximal sets with base size one less than that of the matroid). Prove that M_{24} acts transitively on unordered bases but not on ordered bases.

4.15. Show that the minimal base size of the holomorph of a non-abelian finite simple group is 3.

4.16. Find the minimal base size and the greedy base size of S_n acting on 2-sets.

4.17. Let G be the group $\mathrm{PSL}(2n, q)$, acting on the coset space $H\backslash G$, where $H = \mathrm{PSp}(2n, q)$ (the *projective symplectic group*, see [62, 171]). Prove that, if the degree is n, then $|G| = n^{2+o(1)}$. Deduce that G has no base of size 2. Does it have a base of size 3?

4.18. Construct, for every n, a group of degree $2^n + 2n$ and order 2^n having irredundant bases of all possible cardinalities i such that $1 \leq i \leq n$.

4.19. Show that there is no constant c such that, if a primitive group has minimal base size b, then any irredundant base has size at most cb.

4.20. Prove directly that the bases of a geometric group satisfy the matroid axioms.

4.21. Suppose that G is a permutation group having the property that $G_{\alpha\beta} = 1$ for any $\alpha \neq \beta$. Prove that G is a Frobenius group, and has one orbit on which it has its Frobenius action, all other orbits (if any) being regular.

Hence show that an IBIS group of rank 2, in which every pair of distinct points is a base, is transitive.

Deduce that an IBIS group of rank r, in which every r-tuple of distinct points is a base, is $(r-1)$-transitive.

(Such groups can be classified for $r \geq 3$ without using CFSG. The case $r = 3$ was done by Zassenhaus, Feit, Ito and Suzuki; the groups are known as *Zassenhaus groups*. See Passman [146] for an account of this.)

4.22. (Bochert's bound) Prove that a primitive subgroup of S_n, other than S_n and A_n, has index at least $\lceil n/2 \rceil!$ in S_n.

[*Hint:* Let Γ be a subset of Ω of maximal cardinality subject to $G \cap \mathrm{Sym}(\Gamma) = 1$. Then $|\mathrm{Sym}(\Omega) : G| \geq |\mathrm{Sym}(\Gamma)|$. If $|\Gamma| \geq n/2$, we are done; so suppose that $\Gamma < n/2$. Then there is a non-identity element $g \in G \cap \mathrm{Sym}(\Omega \setminus \Gamma)$. If α is moved by g, there is a non-identity element $h \in G \cap \mathrm{Sym}(\Gamma \cup \{\alpha\})$. Then the supports of g and h meet only in α, and so their commutator is a 3-cycle in G; hence $G \geq \mathrm{Alt}(\Omega)$. (Compare Section 6.1.)]

4.23. This exercise gives an elementary bound, due to Wielandt [185], for the degree of transitivity of a finite permutation group, modulo the truth of

Schreier's Conjecture. Prove that there is no 8-transitive finite permutation group, other than symmetric or alternating groups, as follows. Let G be 8-transitive, and let H be the stabiliser of five points. Then $N_G(H)$ is sharply 5-transitive on the fixed points; that is, $N_G(H)/H \cong S_5$. By Burnside's Theorem, H has a unique minimal normal subgroup T, which is either an elementary abelian 2-group or non-abelian simple. In the second case, the composition factor A_5 of $N_G(H)$ cannot act on T, by the truth of the Schreier Conjecture, so lies in the centraliser of T. Thus, G contains $T \times A_5$, and so contains a 3-cycle. If T is elementary abelian, consider instead the 6-point stabiliser.

Remark. Can you replace 8 by 7 in the above without invoking CFSG (except via the Schreier Conjecture)?

4.24. Prove that, for sufficiently large n, a graph Γ on which the symmetric group S_n acts distance-transitively must be a Johnson graph, an antipodal quotient of a Johnson graph, or an odd graph.

4.25. (a) Let $\mathcal{F}$ be the family of subspaces (flats) of a matroid on X. Take $F_1, F_2 \in \mathcal{F}$ with $F_1 \subset F_2$ and $\rho(F_2) \geq \rho(F_1) + 3$. Consider the incidence structure whose *points* (resp. *lines*) are the subspaces F with $F_1 \subset F \subset F_2$ and $\rho(F) = \rho(F_1) + 1$ (resp. $\rho(F) = \rho(F_1) + 2$), and *incidence* is inclusion. Prove that this structure is a linear space.

(b) Prove that the family $\mathcal{F}$ consists of the flats of a matroid if and only if it satisfies the two conditions at the end of Section 4.15.

4.26. Prove that a primitive permutation group of degree $2p$, where p is an odd prime, is almost simple.

CHAPTER 5

Oligomorphic groups

5.1 The random graph

As we mentioned in Chapter 2, a random (labelled) graph on a given set X of vertices can be chosen by deciding, independently with probability $1/2$, whether each unordered pair of points should be joined by an edge or not.

If $|X| = n$ is finite, then every possible graph on n vertices occurs with positive probability. Moreover, the probability that the random graph is isomorphic to a given graph Γ is inversely proportional to the number of automorphisms of Γ. For the set of all graphs on X which are isomorphic to Γ is an orbit of S_n; and the orbit length is equal to $n!$ divided by the order of the stabiliser of Γ (this stabiliser being just $\mathrm{Aut}(\Gamma)$, as in Section 2.3).

For infinite sets, however, the picture is very different. The following paradoxical result was observed by Erdős and Rényi [66] in 1963.

Theorem 5.1 *There is a countable graph R with the property that, with probability* 1, *a countable random graph is isomorphic to R.*

And we will see that, in sharp contrast to the finite case, R has a very large automorphism group.

Proof. The proof follows immediately from two facts below. First, we say that a graph Γ has *property* $(*)$ if the following is true:

> For any two finite disjoint sets U and V of vertices, there exists a vertex z joined to every vertex in U and to no vertex in V.

We defer for a moment the question whether any graphs satisfying property $(*)$ actually exist.

Fact 1. Any two countably infinite graphs having property $(*)$ are isomorphic.

Proof. Let Γ_1 and Γ_2 be the graphs. We build an isomorphism in stages as follows. Given any isomorphism θ from a finite set A_1 of vertices of Γ_1 to a finite set A_2 of vertices of Γ_2, and a vertex z_1 of Γ_1 not in A_1, we can extend θ to $A_1 \cup \{z_1\}$ as follows. Let U_1 and V_1 be the sets of neighbours and non-neighbours of z_1 in A_1 respectively. Let $U_2 = U_1\theta$ and $V_2 = V_1\theta$. Then by $(*)$, there is a vertex z_2 of Γ_2 joined to everything in U_2 but nothing in V_2. Now we may extend θ to map z_1 to z_2. (Note that this only uses property $(*)$ in Γ_2.)

Enumerate the vertices of Γ_1 and Γ_2, as, say, $(a_1, a_2, \ldots)$ and $(b_1, b_2, \ldots)$ respectively. Start with the empty isomorphism θ_0. Now at an odd-numbered stage n, let z_1 be the first point (in the enumeration of Γ_1) which is not in the domain of θ_{n-1}, and extend (as above) θ_{n-1} to θ_n including z_1 in its domain. At even-numbered stages n, choose the first point z_2 (in the enumeration of Γ_2) not in the range of θ_{n-1}, and (applying the construction in reverse) extend θ_{n-1} to θ_n having z_2 in its range.

The alternation of back and forth ensures that the resulting map θ after all of the finite stages is defined on all of Γ_1 and its image is all of Γ_2. (If we only went forwards, we could not ensure that θ is onto.) So θ is an isomorphism from Γ_1 to Γ_2.

Fact 2. With probability 1, a random countable graph has property $(*)$.

Proof. We have to show that the event that $(*)$ fails has probability zero. We can write this event as the union, over all choices of finite disjoint sets U and V, of the event that $(*)(U,V)$ fails, where $(*)(U,V)$ denotes property $(*)$ for a particular choice of U and V. Now there are only countably many such pairs (U,V); since a countable union of null sets is null (see below), it is enough to prove that the failure of $(*)(U,V)$ has probability 0.

But we can estimate this probability. Let $k = |U \cup V|$, and list the points outside $U \cup V$ as $(z_1, z_2, \ldots)$. The event that z_i is bad (that is, does not fulfil the requirements of $(*)(U,V)$) has probability $1 - 1/2^k$. Since these events, for different i, are independent, the event that the first n points are all bad has probability $(1 - 1/2^k)^n$, which tends to zero as $n \to \infty$; so the event that $(*)(U,V)$ fails is null, as required.

Null sets. In an infinite measure space, a set S is null if, for every $\epsilon > 0$, there is a set S^ϵ of measure less than ϵ containing S. We used above the fact that a countable union of null sets is null. This is shown as follows. Let $S_1, S_2, \ldots$ be null sets with union S, and let a positive number ϵ be given. For each n, choose a set $S_n^{\epsilon/2^n}$ containing S_n and having measure less than $\epsilon/2^n$. The union of the sets $S_n^{\epsilon/2^n}$ contains S and has measure less than $\sum_{n\geq 1} \epsilon/2^n = \epsilon$, as required.

Completion of the proof. Fact 2 shows that a countable graph with property $(*)$ exists (an event with probability 1 must occur). Then Fact 1 shows that all such graphs are isomorphic, and the rest is clear. This is perhaps the ultimate non-constructive existence proof!

The graph R has many further properties, notably the following:

Fact 3. Every finite or countable graph is embeddable in R as an induced subgraph. (We say that R is *universal.*) This is proved by the argument of Fact 1, taking Γ_1 to be an arbitrary graph and $\Gamma_2 = R$, and extending the map θ in the forward direction only. (As we noted there, extending θ in the forward direction only requires property $(*)$ to hold in Γ_2.)

Fact 4. Every isomorphism between finite induced subgraphs of R can be extended to an automorphism of R. We say that R is *homogeneous.* To see this, we again modify the argument of Fact 1, taking $\Gamma_1 = \Gamma_2 = R$, and taking the starting map θ_0 to be the given finite isomorphism instead of the empty map: the final map θ is an isomorphism from R to itself, that is, an automorphism of R.

We will see, when we discuss Fraïssé's Theorem, that the properties of Facts 3 and 4 characterise R as a countable graph: it is *the* countable universal homogeneous graph.

As an application, we show:

Theorem 5.2 $\mathrm{Aut}(R)$ *is a primitive rank* 3 *permutation group on the vertex set of* R.

Proof. Any two 1-vertex subgraphs of R are (trivially) isomorphic; since R is homogeneous, the isomorphism between them extends to an automorphism of R. This means that $\mathrm{Aut}(R)$ is transitive on the vertex set. In a similar way, it has three orbits on ordered pairs of vertices, namely $\{(\alpha,\alpha) : \alpha \in \Omega\}$, $\{(\alpha,\beta) : \alpha \sim \beta\}$, and $\{(\alpha,\beta) : \alpha \not\sim \beta\}$, where we use $\sim$ to denote adjacency and $\not\sim$ for non-adjacency. Thus $\mathrm{Aut}(R)$ has rank 3.

We show primitivity using Higman's Theorem: we have to check that the non-diagonal orbital graphs are connected. The previous paragraph shows that there are just two such graphs, which are R and its complement. The connectedness of R follows from property $(*)$: for, if u_1 and u_2 are any two vertices, then taking $U = \{u_1, u_2\}$ and $V = \emptyset$ we find a common neighbour z of u_1 and u_2. The connectedness of the complement is proved similarly, or we could observe that the complement of R is isomorphic to R (since it is countable and has property $(*)$).

5.1.1 Appendix: A construction of R

We have so far given only a non-constructive existence proof for the graph R. For some purposes, it may be more satisfactory to have an explicit construction. The first such construction was given by Richard Rado [153] in 1964. Several other constructions are known; two of the more unusual are given as exercises at the end of this chapter.

After my EIDMA lecture on this topic, two students on the course, Jürgen Müller and Max Neunhöffer, produced the following construction. With their permission, I have included it in their own words.

Theorem 5.3 *The following construction produces a countable graph R with the property $(*)$: for any finite disjoint sets U and V of vertices there is a vertex z joined to everything in U and nothing in V.*

Let the vertices of R be enumerated by $v_0, v_1, \ldots$. Consider the (unique) triadic representation of j:

$$j = \sum_{k=0}^{n-1} a_k 3^k \quad \text{with } a_k \in \{0,1,2\}\ \forall k \quad \text{and } a_{n-1} \neq 0.$$

Then the vertex v_j is joined exactly to those v_i with $i < j$ for which $a_i = 1$ in the above representation.

Proof. We give an algorithm for constructing R which shows that it satisfies property $(*)$. Then the above formula is verified.

The graph R is determined if we specify, for each vertex, to which of the vertices of lower index it is joined. The algorithm produces R inductively. There is no information needed for v_0. So we have already one vertex. Now we add new vertices (with certain edges to the vertices with lower index), such that the property $(*)$ is satisfied for U and V, which are subsets of $\{0,1,2,\ldots,n-1\}$ where n runs inductively through the natural numbers, starting with 1. To enumerate all possibilities of choosing U and V disjoint in $\{0,1,2,\ldots,n-1\}$ we look at the triadic representations $(a_{n-1}, a_{n-2}, \ldots, a_1, a_0)$ of the $2\cdot 3^{n-1}$ numbers from 3^{n-1} to $3^n - 1$, which all start with 1 or 2. $a_i = 1$ in this representation means that $i \in U$, $a_i = 2$ means $i \in V$ and $a_i = 0$ means that i is not in U or V. We only need those configurations where $n-1$ lies in U or in V. For k running through these numbers we add a new vertex, joined to exactly those vertices v_i for which $a_i = 1$. So the property $(*)$ is satisfied for U and V in $\{0,1,2,\ldots,n-1\}$. As n grows arbitrarily large, the property $(*)$ is satisfied in R, because every (U,V) of disjoint finite subsets of the natural numbers is contained in some $\{0,1,2,\ldots,n-1\}$.

This algorithm produces exactly the graph which is described in the above formula, because if $j = \sum_{k=0}^{n-1} a_k 3^k$ with $a_k \in \{0,1,2\}$ and $a_{n-1} \neq 0$ the algorithm describes the edges from v_j to v_i with $i < j$ exactly in the nth step, when we look at the triadic expansion of j.

5.2 Oligomorphic groups

The permutation group G on Ω is *oligomorphic* if the number $f_n(G)$ of orbits of G on the set of n-element subsets of Ω is finite for all $n \in \mathbb{N}$, or equivalently, if the number $F_n(G)$ of orbits of G on the set of ordered n-tuples of distinct elements of Ω is finite for all $n \in \mathbb{N}$. The equivalence of these two definitions follows from the fact that

$$f_n(G) \leq F_n(G) \leq n!f_n(G),$$

which holds because each orbit of G on n-sets corresponds to between 1 and $n!$ orbits on n-tuples.

Now $F_n(G) = 1$ if and only if G is n-transitive. We say that G is *highly transitive* if it is n-transitive for all $n \in \mathbb{N}$. Similarly, we say that G is *n-set-transitive* if $f_n(G) = 1$, and *highly set-transitive* if it is n-set-transitive for all $n \in \mathbb{N}$. (The terms 'n-homogeneous' and 'highly homogeneous' are often used for what is here called 'n-set-transitive' and 'highly set-transitive'; but, as we saw in the last section, we want to reserve the term 'homogeneous' for a different use.)

Example. The group $G = \mathrm{Aut}(R)$ is oligomorphic, where R denotes the random graph. Two n-tuples of distinct elements lie in the same orbit if and only if the natural map between them is an isomorphism of induced subgraphs (since R is homogeneous), that is, if and only if they are the same labelled graph. So $F_n(\mathrm{Aut}(R))$ is equal to the number of labelled graphs on n vertices, which is $2^{n(n-1)/2}$. Similarly, two n-sets lie in the same orbit if and only if the induced subgraphs are isomorphic (that is, the same unlabelled graph); so $f_n(G)$ is the number of unlabelled graphs on n vertices.

As this example suggests, calculating $f_n(G)$ and $F_n(G)$ for oligomorphic groups G is closely related to enumeration problems for unlabelled and labelled combinatorial structures of various kinds. The term *oligomorphic* (literally 'few shapes') is intended to suggest that there are only finitely many shapes of finite subsets (where a 'shape' is a G-orbit). The only other use of this word that I know is to describe a kind of computer virus which can change its form but only among a limited set of possibilities.

The *main problem* on oligomorphic groups is:

> How do the sequences $(f_n(G))$ and $(F_n(G))$ of positive integers behave? In particular, do they grow rapidly? smoothly?

5.3 First-order logic

It is not possible to give a complete course on first-order logic in this short section: only the most basic ideas are introduced, and many details are ignored. A very complete general reference is the book by Hodges [93].

A *first-order language* consists of various logical symbols (connectives, quantifiers, punctuation, and names for variables) together with some non-logical symbols adapted for the particular application in mind: these are function symbols, relation symbols, and constant symbols. Each function symbol or relation symbol comes equipped with an *arity*, the number of arguments it takes. The terms and formulae of the language are built up by recursive rules.

If $\mathcal{L}$ is a first-order language, a *structure* over $\mathcal{L}$, or *interpretation* of $\mathcal{L}$, is a non-empty set X on which there are some functions, relations and constants corresponding to the symbols of the language. Thus, to each n-ary function symbol f corresponds a function $f : X^n \to X$; to each n-ary relation symbol R corresponds an n-ary relation on X (a subset of X^n); and to each constant symbol corresponds an element of X. Now each formula in the language has a 'meaning' in the interpretation. It may be true or false; in general, this depends on the values assigned to the variables of the language (these values are not specified by the interpretation). In the particular case where the formula has no free occurrences of variables (that is, not bound by quantifiers), its truth value is independent of the values of the variables, and depends only on the interpretation. A formula with no free occurrences of variables is called a *sentence*; a structure in which the sentence σ (or set Σ of sentences) is true is called a *model* of σ (or Σ).

For example, if we wanted to do group theory, we might take a language with a binary function symbol m (for the group operation), a unary function symbol i (for inversion), and a constant symbol e (for the identity). The group axioms are sentences, viz.

$$(\forall x)(\forall y)(\forall z)(m(m(x,y),z) = m(x,m(y,z))),$$
$$(\forall x)((m(x,e) = x) \wedge (m(e,x) = x)),$$
$$(\forall x)((m(x,i(x)) = e) \wedge (m(i(x),x) = e)).$$

Now a structure for the language is a set G carrying a binary function, a unary function, and containing a distinguished constant; it is a group if and only if it is a model for the three sentences above.

Again, if we wanted to do graph theory, we could take a language with a binary relation symbol $\sim$; a (loopless undirected simple) graph is a structure over this language which is a model for the sentences

$$(\forall x)(\neg(x \sim x)),$$
$$(\forall x)(\forall y)((x \sim y) \to (y \sim x)).$$

We say that a language or structure is *relational* if it contains no function or constant symbols. The important observation is that any subset of a relational structure carries an *induced substructure* obtained by restricting all relations to the subset. On the other hand, a subset of a general first-order structure carries a substructure if and only if it contains all the constants and is closed under all the functions. (Thus, every subset of a graph carries a subgraph, but not every subset of a group carries a subgroup!)

In Section 4.12, we used the *Compactness Theorem*:

Theorem 5.4 *Let Σ be a set of first-order sentences. Suppose that every finite subset of Σ has a model. Then Σ has a model.*

The specific use we made of this theorem was as follows.

Theorem 5.5 *Let Σ be a set of first-order sentences. Suppose that Σ has arbitrarily large finite models. Then Σ has an infinite model.*

Proof. For every n, there is a sentence τ_n whose interpretation is that there are at least n elements in the domain. (This sentence asserts the existence of n elements which are pairwise unequal.) Let $\mathrm{T} = \{\tau_n : n \in \mathbb{N}\}$. Then, by assumption, every finite subset of $\Sigma \cup \mathrm{T}$ has a model; so the whole set has a model, which is necessarily an infinite model of Σ.

We conclude this section with a technical result about the random graph which we need later. First, we write down some sentences which capture property (*) of a graph. For any positive integers m and n, let $\phi_{m,n}$ be the sentence asserting that, given any vertices $u_1, \ldots, u_m$, $v_1, \ldots, v_n$, all distinct, there is a vertex z which is joined to all the u_i and none of the v_j. This is cumbersome to write down. The statement that the us and vs are all distinct is the conjunction of $(m+n)(m+n-1)/2$ formulae of the shape $(\neg(x = y))$, where x and y are us or vs. However, in principle this can be done. Now Fact 1 from Section 5.1 shows that any countable structure over this language which satisfies the two graph axioms given above and all the sentences $\phi_{m,n}$ is isomorphic to the random graph R. It follows from general results in logic that any first-order sentence which is true in R can be deduced from the graph axioms and the sentences $\phi_{m,n}$, in a suitable formal deduction system for first-order logic. (This is the Gödel–Henkin Completeness Theorem.)

Using this, we can prove the following result due to Glebskii *et al.* [80], Fagin [70] and Blass and Harary [21]. We say that *almost all finite graphs have property* P if the proportion of (labelled) n-vertex graphs having property P tends to 1 as $n \to \infty$.

Theorem 5.6 *Let σ be a sentence in the first-order language of graph theory. Then σ holds in almost all finite graphs if and only if σ holds in R. In particular, either σ holds in almost all finite graphs, or its negation does.*

Proof. Suppose first that σ is true in R. Then σ can be deduced from the graph axioms and the sentences $\phi_{m,n}$. Since deductions are finite, σ can be deduced from a finite set of these sentences. (This argument is the Compactness Theorem.) So it will suffice to show that each sentence $\phi_{m,n}$ holds in almost all finite graphs. We can estimate this number: the probability that $\phi_{m,n}$ fails in a random N-vertex graph is not greater than

$$\binom{N}{m+n}\left(1-2^{-(m+n)}\right)^{N-m-n},$$

this being the product of the number of choices of a set of $m+n$ vertices and the probability that no vertex outside is correctly joined. This number tends to 0 as $N \to \infty$.

For the converse, suppose that σ is not true in R. Then $(\neg\sigma)$ is true in R, and hence in almost all finite graphs (by what we have just proved).

This means that there is a *zero–one law* for first-order properties of finite graphs: the probability of any such property tends to either zero or one.

5.4 Automorphism groups and topology

There is a natural topology on the symmetric group $\mathrm{Sym}(\Omega)$, the topology of *pointwise convergence*. For simplicity we describe it only in the case where Ω is countable, since in general a more elaborate notion of 'convergence' is required.

Suppose that $\Omega = \mathbb{N}$. A sequence $(g_1, g_2, \ldots)$ of permutations *converges* to the permutation g if, for any n, there exists m such that $ng_i = ng$ for all $i \geq m$. That is, the permutations in the sequence agree on longer and longer initial segments of $\mathbb{N}$, and the resulting function is the permutation g. It is possible to derive the topology from a metric, though we don't need this fact.

The topology gives us a convenient reduction for oligomorphic groups, since the closure of a subgroup is the set of all permutations which are limits of sequences of elements of the subgroup.

Theorem 5.7 *Let G be a permutation group on Ω. Then the closure of G in* $\mathrm{Sym}(\Omega)$ *is the group of all permutations of Ω which preserve all the orbits of G on Ω^n for all n: that is, the largest group having the same orbits on Ω^n as G.*

In particular, if G is oligomorphic, and we are interested in the orbit-counting sequences, then we lose nothing by assuming that G is closed. Now the closed groups are described in the next result. A permutation g is an *automorphism* of a first-order structure M on Ω if it preserves all the relations, commutes with the functions (in the sense that $f(x_1g, \ldots, x_ng) = f(x_1, \ldots, x_n)g$), and fixes all the constants.

Theorem 5.8 *The following properties of a permutation group G on Ω are equivalent if Ω is countable:*

(a) G is closed in $\mathrm{Sym}(\Omega)$*;*

(b) $G = \mathrm{Aut}(M)$ for some first-order structure M on Ω;

(c) $G = \mathrm{Aut}(M)$ for some relational structure M on Ω;

(d) $G = \mathrm{Aut}(M)$ for some homomgeneous relational structure M on Ω.

Proof. (b) implies (a): clearly a pointwise limit of automorphisms is an automorphism.

(d) implies (c) implies (b): trivial.

(a) implies (d): Take any closed group G. For each orbit O_i of G on Ω^n, we put into the language an n-ary relational symbol R_i which is true just for the n-tuples of O_i in the interpretation. We do this for every n. Now clearly G is contained in the automorphism group of this relational structure. We show that every automorphism lies in the closure of G. Let g be an automorphism. Given n, there is a relation R_i satisfied by $(1, 2, \ldots, n)$, and hence also by their images under g. So $(1, 2, \ldots, n)$ and $(1g, 2g, \ldots, ng)$ lie in the same G-orbit, and there exists $g_n \in G$ with $xg_n = xg$ for $x \leq n$. Then the sequence (g_n) converges to g. Since G is closed, $g \in G$.

Finally, we have to show that this relational structure is homogeneous. Suppose that there is an isomorphism f between finite substructures, taking x_i to y_i for $1 \leq i \leq n$. Then the n-tuples $(x_1, \ldots, x_n)$ and $(y_1, \ldots, y_n)$ satisfy the same relation, and hence lie in the same G-orbit, whence there is an element of G extending f.

We see that the situation in the random graph, where counting orbits of the automorphism group on ordered or unordered n-tuples is the same as counting labelled or unlabelled graphs, is typical for oligomorphic groups.

5.5 Countably categorical structures

It is possible to say everything about a finite first-order structure in a collection of first-order sentences (that is, to specify the structure up to isomorphism). However, according to the Löwenheim–Skolem theorems, this is not possible for an infinite structure: if a set of sentences has an infinite model, then it has arbitrarily large countable models. The best we can hope to do is to specify it if the additional assumption that the structure is countable is given to us.

Accordingly, we say that a countable first-order structure M is *countably categorical* or *$\aleph_0$-categorical* if any countable structure N over the same language which satisfies the same first-order sentences is isomorphic to M. In other words, we can specify M by first-order axioms together with the assumption of countability. (Note that logicians more usually apply this term to the set of sentences rather than to the structure.)

An example of a countably categorical structure is the ordered set $\mathbb{Q}$ of rational numbers. This is the content of *Cantor's Theorem:*

Theorem 5.9 *A countable dense totally ordered set without least or greatest element is isomorphic to* $\mathbb{Q}$.

For all the conditions of the theorem, except countability, can be expressed as first-order sentences.

Another example is the random graph R: we saw that it is characterised as a countable structure by the graph axioms and the sentences $\phi_{m,n}$.

The following important characterisation of countably categorical structures (and the reason for their appearance in this chapter) was found independently by Engeler [65], Ryll-Nardzewski [157] and Svenonius [170] in 1959.

Theorem 5.10 *The countable first-order structure M is countably categorical if and only if* $\mathrm{Aut}(M)$ *is oligomorphic.*

In the course of the proof, the following is shown. We will need this later.

Theorem 5.11 *Let M be a countable and countably categorical first-order structure. Then two n-tuples of elements of M lie in the same orbit of* $\mathrm{Aut}(M)$ *if and only if they satisfy the same n-variable first-order formulae.*

We know that the ordered set $\mathbb{Q}$ is countably categorical. Let us check that its automorphism group is oligomorphic. In fact, as we will see, it is highly set-transitive.

Given two n-tuples $(a_1, \ldots, a_n)$ and $(b_1, \ldots, b_n)$ of distinct rationals, arranged in ascending order, we want to carry the first to the second by a permutation of $\mathbb{Q}$ which preserves the order (see Figure 5.1). For each $i \leq n-1$, the interval $[a_i, a_{i+1}]$ can be mapped in order-preserving fashion to $[b_i, b_{i+1}]$ by a linear map

$$x \mapsto b_i + \frac{(b_{i+1} - b_i)(x - a_i)}{(a_{i+1} - a_i)}.$$

For numbers less than a_1, we apply a translation

$$x \mapsto x + b_1 - a_1,$$

and similarly for numbers greater than a_n. Putting these pieces together, we obtain a piecewise-linear order-preserving map taking the first n-tuple to the second.

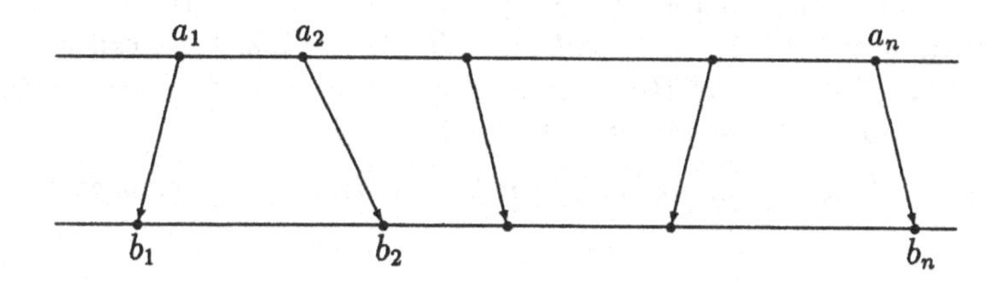

Figure 5.1: Order-automorphism of $\mathbb{Q}$

Remark. The piecewise-linear order-preserving permutations form a group which is countable and highly set-transitive. The closure of this group is the group of all order-preserving permutations (which is uncountable).

5.6 Homogeneous structures

All structures in this section are relational. We make some definitions.

- The *induced substructure* of the structure M on a subset Y consists of Y with all the relations $R \cap Y^n$, for each n and each n-ary relation R of M.
- The structure M is *homogeneous* if any isomorphism between finite substructures of M extends to an automorphism of M.
- The *age* of M is the class of all finite structures (over the same language) which can be embedded in M.

The term 'age' is due to Fraïssé [77], along with other colourful terminology. For example, Fraïssé speaks of a 'relation', rather than a 'relational structure'; then he can say that one relation is 'older than' another. He also has the concept of a 'rich relation'.

Fraïssé [76] was able to characterise the ages of homogeneous relational structures as follows.

Theorem 5.12 *Let C be a class of finite structures over a relational language. Then C is the age of a countable homogeneous relational structure M if and only if the following four conditions hold:*

(a) C is closed under isomorphism;

(b) C is closed under taking induced substructures;

(c) C contains only countably many non-isomorphic structures;

(d) *$\mathcal{C}$ has the* amalgamation property; *that is, given $B_1, B_2 \in \mathcal{C}$ and an isomorphism $f : A_1 \to A_2$ with $A_i \subseteq B_i$ for $i = 1, 2$, there exists $C \in \mathcal{C}$ in which both B_1 and B_2 are embedded so that A_1 and A_2 are identified according to the isomorphism f (see Figure 5.2).*

Moreover, if these conditions hold, then M is unique up to isomorphism.

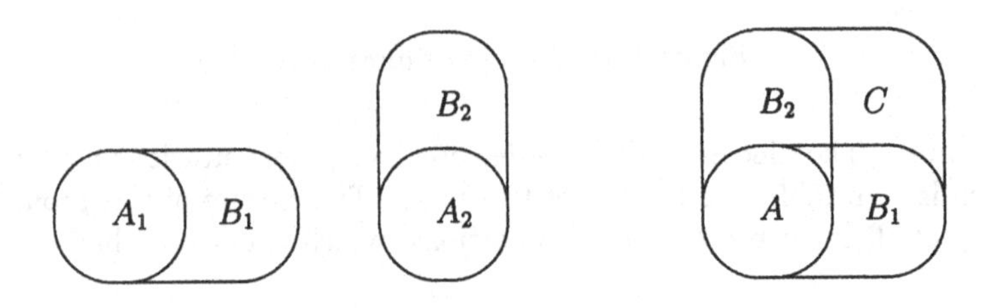

Figure 5.2: The amalgamation property

Note that, in the amalgamation property, the intersection of B_1 and B_2 inside C is permitted to be larger than A.

A class $\mathcal{C}$ of finite structures satisfying the hypotheses of this theorem is called a *Fraïssé class*, and the structure M guaranteed by the theorem is its *Fraïssé limit.*

Examples.

- $\mathcal{C}$ is the class of finite graphs; M is the random graph R.
- $\mathcal{C}$ is the class of finite totally ordered sets; M is the ordered set $\mathbb{Q}$.
- A *k-uniform hypergraph* consists of a set of *vertices* equipped with a collection of k-element subsets called *hyperedges.* The class of all finite k-uniform hypergraphs is a Fraïssé class; its Fraïssé limit is the *random k-uniform hypergraph*, whose automorphism group is $(k-1)$-transitive but not k-transitive. Thus, every possible degree of transitivity is realised by an infinite permutation group.

Remark. Let M be a homogeneous structure, $\mathcal{C}$ its age, and G its automorphism group. Then

- $f_n(G)$ is the number of unlabelled n-element structures in $\mathcal{C}$;
- $F_n(G)$ is the number of labelled n-element structures in $\mathcal{C}$.

So the orbit-counting sequences solve combinatorial enumeration problems for Fraïssé classes.

5.7 Cycle index

It is natural, when given a sequence of natural numbers, to define a generating function for it. Now it turns out that suitable generating functions for both the sequences $(f_n(G))$ and $(F_n(G))$ can be obtained as specialisations of a power series in infinitely many indeterminates, related to the 'cycle index' of the theory of enumeration under group action developed by Redfield, Pólya and de Bruijn.

The definitions are as follows.

- Let g be a permutation on Ω, with $|\Omega| = n$. Let $c_k(g)$ be the number of k-cycles in the cycle decomposition of g. Then we put

$$z(g) = s_1^{c_1(g)} s_2^{c_2(g)} \cdots s_n^{c_n(g)},$$

  where $s_1, s_2, \ldots$ are indeterminates.

- Let G be a permutation group on Ω, with $|\Omega| = n$. Then we put

$$Z(G) = \frac{1}{|G|} \sum_{g \in G} z(g).$$

- There is no hope of defining $Z(G)$ for infinite groups G. Instead, if G is an oligomorphic group on an arbitrary set Ω, we put

$$\tilde{Z}(G) = \sum_{\Delta} Z(G_{\Delta}^{\Delta}),$$

  where the sum is over a set of orbit representatives Δ for the action of G on the finite subsets of Ω, and G_{Δ}^{Δ} is the group induced on Δ by its setwise stabiliser in G. The sum is well-defined: for, given a monomial $s_1^{a_1} \cdots s_r^{a_r}$, the only terms in the sum which contribute to its coefficient are those arising from sets Δ of cardinality $a_1 + 2a_2 + \cdots + ra_r$; and there are only finitely many such sets Δ (as G is oligomorphic). We adopt the convention that the empty set Δ contributes a term 1 to the sum.

Remark. Surprisingly, in the case of finite groups, we get nothing new: it can be shown that, if G is a finite permutation group, then $\tilde{Z}(G) = Z(G; s_i \leftarrow s_i + 1)$, where we use the notation $F(s_i \leftarrow x_i)$ for the result of substituting x_i for s_i for all i. However, this substitution is not valid for formal power series in general: the expressions which we substitute for the s_i must have zero constant term. So for infinite oligomorphic groups we have produced something genuinely new.

Example. For the symmetric group S_3 of degree 3, we have

$$\begin{aligned}\tilde{Z}(S_3) &= 1 + s_1 + \tfrac{1}{2}(s_1^2 + s_2) + \tfrac{1}{6}(s_1^3 + 3s_1s_2 + 2s_3)\\ &= \tfrac{1}{6}((s_1+1)^3 + 3(s_1+1)(s_2+1) + 2(s_3+1))\\ &= Z(S_3; s_i \leftarrow s_i + 1).\end{aligned}$$

Now we use the following 'generating functions' for the orbit-counting sequences:

- the *ordinary generating function*

$$f_G(t) = \sum_{n\geq 0} f_n(G)t^n$$

  for $(f_n(G))$;

- the *exponential generating function*

$$F_G(t) = \sum_{n\geq 0} \frac{F_n(G)t^n}{n!}$$

  for $(F_n(G))$.

This accords with standard practice in combinatorial enumeration theory, where ordinary and exponential generating functions are used for counting unlabelled and labelled structures respectively. But more important, these generating functions are specialisations of the modified cycle index just defined:

Theorem 5.13 *For any oligomorphic group G,*

$$\begin{aligned}F_G(t) &= \tilde{Z}(G; s_1 \leftarrow t, s_i \leftarrow 0 \text{ for } i > 1),\\ f_G(t) &= \tilde{Z}(G; s_i \leftarrow t^i \text{ for all } i).\end{aligned}$$

Example. Let $G = A = \mathrm{Aut}(\mathbb{Q}, <)$ (the group of order-preserving permutations of $\mathbb{Q}$). We saw that this group is highly set-transitive, and the group induced on any finite set by its setwise stabiliser is trivial (with cycle index s_1^n for an n-set). So

$$\tilde{Z}(A) = \sum_{n\geq 0} s_1^n = \frac{1}{1 - s_1}.$$

On specialising we get $F_A(t) = f_A(t) = 1/(1-t)$, and so $f_n(A) = 1$, $F_n(A) = n!$, for all n.

Example. A calculation given in Section 5.13 shows that, if S denotes the symmetric group of countable degree, then

$$\tilde{Z}(S) = \exp\left(\sum_{n\geq 1} \frac{s_n}{n}\right).$$

Hence we obtain

- $F_S(t) = \exp(t) = \sum_{n\geq 0} t^n/n!$, whence $F_n(S) = 1$ for all n;
- $f_S(t) = \exp(\sum t^n/n) = \exp(-\log(1-t)) = 1/(1-t) = \sum_{n\geq 1} t^n$, whence $f_n(S) = 1$ for all n.

The behaviour of the modified cycle index with respect to direct and wreath products can also be described. We take the intransitive action of the direct product (on the disjoint union of the sets), and the imprimitive action of the wreath product. Also, we take the action of the stabiliser of a point α on $\Omega \setminus \{\alpha\}$.

Theorem 5.14 *(a)* $\tilde{Z}(G \times H) = \tilde{Z}(G) \cdot \tilde{Z}(H)$.

(b) $\tilde{Z}(G \operatorname{Wr} H) = \tilde{Z}(H; s_i \leftarrow \tilde{Z}(G; s_j \leftarrow s_{ij}) - 1)$.

(c) $\tilde{Z}(G_\alpha) = \partial\tilde{Z}(G)/\partial s_1$ *if G is transitive.*

In (c), if G is not transitive, then replace the left-hand side by the sum of the modified cycle indices of point stabilisers, summed over a set of orbit representatives.

The modified cycle indices of $G \times H$ and $G \operatorname{Wr} H$ in their product actions are not determined by those of the factors. Even in the very simplest case, that of $S \times S$ in its product action on $\Omega \times \Omega$, determination of the numbers f_n is an unsolved combinatorial problem! (These are the numbers of zero–one matrices with n non-zero entries and no zero rows or columns, up to row and column permutations.)

5.8 A graded algebra

A *graded algebra*, or $\mathbb{N}$-*graded algebra*, is an algebra A over a field F which is a direct sum of F-vector spaces V_n for $n \in \mathbb{N}$, with the property that $V_n \cdot V_m \subseteq V_{n+m}$. If all these spaces are finite-dimensional, the *Poincaré series* of the algebra is the ordinary generating function for their dimensions, that is, $\sum_{n\geq 0}(\dim V_n)t^n$. This condition holds if A is finitely generated. Graded algebras arise in areas such as invariant theory and algebraic geometry.

It is possible to construct a graded algebra whose structure reflects the orbit structure of the oligomorphic group G, in the sense that the dimension of the nth homogeneous component of the algebra is $f_n(G)$ (so that its Poincaré series is $f_G(t)$).

Let Ω be an infinite set, and let $\binom{\Omega}{n}$ denote the set of n-element subsets of Ω. We first define an algebra $\mathcal{A} = \bigoplus_{n\geq 0} V_n$ as follows.

- V_n is the vector space (over $\mathbb{C}$) of functions from $\binom{\Omega}{n}$ to $\mathbb{C}$, with pointwise operations. Then
$$\mathcal{A} = \bigoplus_{n\geq 0} V_n.$$
- For $f \in V_n$, $g \in V_m$, we define fg to be the function in V_{m+n} whose value on the $(m+n)$-set X is given by
$$(fg)(X) = \sum_{\substack{Y\subseteq X\\ |Y|=n}} f(Y)g(X \setminus Y).$$

Then $\mathcal{A}$ is a commutative and associative graded algebra. It is sometimes referred to as the *reduced incidence algebra* of finite subsets of Ω (Rota [156]).

Now let G be a permutation group on Ω. Let V_n^G be the subspace of V_n consisting of functions fixed by G. Since a function is fixed by G if and only if it is constant on all the orbits of G, we see that $\dim V_n^G = f_n(G)$ (if this number is finite). Then put
$$\mathcal{A}^G = \bigoplus_{n\geq 0} V_n^G,$$
the set of fixed elements of G in $\mathcal{A}$.

The structure of the algebra $\mathcal{A}^G$ is known in some cases, especially those where G is the automorphism group of a suitable kind of homogeneous relational structure. Among other results, I mention that
$$\mathcal{A}^{G_1\times G_2} = \mathcal{A}^{G_1} \otimes_{\mathbb{C}} \mathcal{A}^{G_2},$$
where the product $G_1 \times G_2$ has its intransitive action.

Conjecture. If G has no finite orbits, then $\mathcal{A}^G$ is an integral domain.

Remark. The truth of this conjecture would imply that
$$f_{m+n}(G) \geq f_m(G) + f_n(G) - 1$$
for any oligomorphic group G with no finite orbits.

5.9 Monotonicity

It is trivial that $F_n(G) \leq F_{n+1}(G)$ for an infinite oligomorphic group G. For there is a map from the set of $(n+1)$-tuples of distinct elements to the set of n-tuples of distinct elements, given by taking the first n components of the $(n+1)$-tuple; and it maps $(n+1)$-tuples in the same orbit to n-tuples in the same orbit, so it yields a map on orbits, which is clearly onto. Moreover, equality implies that all the $(n+1)$-tuples extending a given n-tuple belong to the same orbit, that is, the stabiliser of any n-tuple is transitive on the remaining points. This implies, after an easy induction, that the group G is $(n+1)$-transitive. Thus, $F_n(G) = F_{n+1}(G)$ implies $F_{n+1}(G) = 1$.

It is also true, but much deeper, that the sequence $(f_n(G))$ is monotonic. Two entirely different proofs of this fact are known. A very brief outline of each is given.

Theorem 5.15 *If G acts on an infinite set Ω, then $f_n(G) \leq f_{n+1}(G)$.*

First proof. This is based on the ideas of Section 5.8. Let e be the constant function in V_1^G with value 1. By combinatorial methods it can be shown that e is not a zero-divisor in the algebra $\mathcal{A}^G$, that is, the map from V_n^G to V_{n+1}^G given by multiplication by e is a monomorphism. So

$$f_{n+1}(G) = \dim V_{n+1}^G \geq \dim V_n^G = f_n(G).$$

Remark. The conjecture in the last section has been strengthened to the conjecture that $\mathcal{A}^G/e\mathcal{A}^G$ is an integral domain if G has no finite orbits. This would then imply that

$$f_{m+n}(G) - f_{m+n-1}(G) \geq (f_m(G) - f_{m-1}(G)) + (f_n(G) - f_{n-1}(G)) - 1$$

for an oligomorphic group G with no finite orbits.

Second proof. This proof uses an extension of Ramsey's Theorem. Suppose that the n-subsets of an infinite set Ω are coloured with r distinct colours $c_1, \ldots, c_r$. By Ramsey's Theorem, there is a colour (say c_1, after re-ordering the colours) and an $(n+1)$-set Y_1 all of whose n-subsets have colour c_1. The extension asserts that, after suitable re-ordering of the colours, there are $(n+1)$-sets $Y_1, , \ldots, Y_r$, such that Y_i contains at least one n-subset of colour i but none of colour j for $j > i$.

There is a 'finite' version of this theorem, but the corresponding 'Ramsey numbers' (viz., how large does Ω have to be in terms of r and n for the conclusion to hold) have not been investigated.

Now choose the colouring so that each colour class is an orbit of G on n-sets (with $r = f_n(G)$). Then the sets $Y_1, \ldots, Y_r$ all lie in different orbits of G; so $f_{n+1}(G) \geq r$.

Both these proofs can be found in [34].

Corollary 5.16 *An $(n+1)$-set-transitive infinite permutation group is n-set-transitive.*

When does equality hold in Theorem 5.15? This is much more difficult to resolve than for the ordered version discussed at the start of the section. There are infinitely many intransitive or imprimitive examples, but these can be 'described' in a fairly precise way. It is conjectured that there is an absolute upper bound for n such that there is a primitive group G such that $f_n(G) = f_{n+1}(G)$ but G is not $(n+1)$-set-transitive, but this conjecture has evaded proof so far. (The largest known value is $n = 6$; the example will be described briefly in the next section.)

5.10 Set-transitive groups

We saw that the group A of order-preserving permutations of the rationals is highly set-transitive. This group is not even 2-transitive, since two distinct points cannot be interchanged. Various subgroups of A (such as the piecewise linear maps, the maps of bounded support, etc.) are also highly set-transitive; but this will not concern us here, where the aim is to describe the closed groups.

An example of a highly set-transitive group which is 2-transitive is obtained by allowing also permutations which reverse the order. This group B contains A as a normal subgroup of index 2. It is not 3-transitive since, of any three points, one comes between the other two, and this *betweenness* relation is preserved by B.

Another approach is to bend the line round into a circle, and consider the group C of permutations which preserve the cyclic order on the complex roots of unity. Again C is not 3-transitive, since it preserves the ternary *circular order* relation consisting of the triples of points in anticlockwise order.

We can combine the two methods and define the group D of permutations preserving or reversing the cyclic order. This group is 3-transitive, but not 4-transitive, since of the three ways of dividing four points into two sets of two, in only one case does each pair separate the other, and this *separation* relation is preserved.

It turns out that no amount of cleverness will yield further examples of highly transitive groups, as *Cameron's Theorem* [31] shows.

Theorem 5.17 *Let G be a highly set-transitive but not highly transitive infinite permutation group on Ω. Then there is a linear or circular order on Ω which is preserved or reversed by G.*

Corollary 5.18 *The only closed subgroups of $S = \mathrm{Sym}(\mathbb{Q})$ containing A are the groups A, B, C, D and S.*

To give a hint at the ideas in the proof, here is a related result, due to John McDermott (unpublished).

Theorem 5.19 *An infinite permutation group which is 3-set-transitive but not 2-transitive preserves a linear order.*

Proof. Let G be such a group. Then G is 2-set-transitive, by Theorem 5.15. So G has just two orbits on ordered pairs of distinct elements; if one of these is Δ, then the other is its converse Δ^*. Represent $(\alpha, \beta) \in \Delta$ by an arrow from α to β.

There are only two possible shapes for a 3-set, as shown in Figure 5.3. Since G is 3-set-transitive, only one of them can occur. But not every set

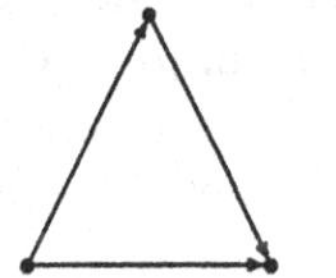

Figure 5.3: 3-sets

can be of the cyclic type: for, given any point α, either two or more arrows leave α, or two or more enter. So every 3-set is of the transitive type, and Δ, regarded as a relation on Ω, is irreflexive, antisymmetric, and transitive; that is, a linear order.

This argument suggests the existence of a function f such that if a group is $f(n)$-set-transitive but not n-transitive then it preserves or reverses a linear or circular order. This was proved in [31]. Subsequently Macpherson [129] found a result close to best possible:

Theorem 5.20 *An $(n+3)$-set-transitive group which is not n-transitive preserves or reverses a linear or circular order.*

In the other direction, the following examples are known:

Theorem 5.21 *(a) For every $n \geq 2$, there is a group which is n-set-transitive but not n-transitive and does not preserve or reverse a linear or circular order.*

(b) There is a group which is 5-set-transitive but not 4-transitive and does not preserve or reverse a linear or circular order.

These examples are constructed using Fraïssé's Theorem. For (a), we take the Fraïssé class C consisting of finite sets with an n-ary relation R satisfying the conditions

- if $R(x_1, \ldots, x_n)$, then $x_1, \ldots, x_n$ are all distinct;
- if $x_1, \ldots, x_n$ are distinct then $R(x_{1\pi}, \ldots, x_{n\pi})$ holds for exactly one $\pi \in S_n$.

Since all n-element structures are isomorphic, the automorphism group of the Fraïssé limit M is n-set-transitive, but clearly not n-transitive.

Part (b) is more interesting. A *boron tree* is defined to be a finite tree in which all vertices have valency 1 or 3. (This represents a hypothetical hydrocarbon in which tetravalent carbon atoms are replaced by trivalent boron.) The leaves (or hydrogen atoms) of a boron tree carry a quaternary relation: of the three ways of dividing four leaves into two sets of two, just one has the property that each pair fails to separate the other. (This is the opposite of what happens in the separation relation of a circular order. See Figure 5.4 below, which shows the boron trees with at most five leaves.)

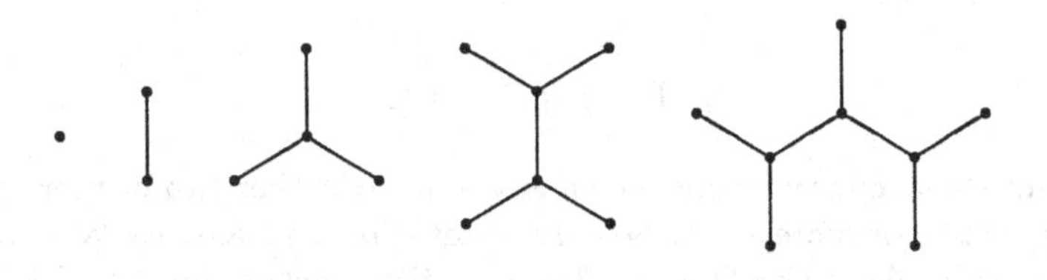

Figure 5.4: Boron trees

It can be shown that the class of finite quaternary relational structures derived from boron trees in this way is a Fraïssé class. The automorphism group G of the Fraïssé limit is clearly 5-set-transitive (since there is only one boron tree with 5 leaves) but not 4-transitive.

In fact, there are only two boron trees with six leaves, and two with seven leaves. So this group satisfies $f_6(G) = f_7(G) = 2$. This is the largest value of n for which $f_n(G) = f_{n+1}(G) > 1$ is known to occur in a primitive group. (See the comments in the last section.)

5.11 Growth rates

For any oligomorphic group G, we know now that the sequence $(f_n(G))$ is non-decreasing, and we understand the groups in which it is 1 for all n. In any other case, it begins to grow at some point. Can anything be said about how fast it grows? Much is known, and yet much remains to be learnt. The most striking theorem is due to Macpherson [127]:

Theorem 5.22 (Macpherson's Theorem) *There is a constant $c > 1$ such that, if G is primitive but not highly set-transitive, then $f_n(G) \geq c^n$ for all sufficiently large n.*

In fact, he shows that we can take $c = \sqrt[5]{2} - \epsilon$ for any $\epsilon > 0$. The slowest known growth rate in primitive groups is $f_n(G) \sim 2^{n-1}/n$.

Here is a very brief sketch of the proof of Macpherson's Theorem. We say that the oligomorphic group G is in the class $\mathcal{E}(c)$ (where c is a real number greater than 1) if $f_n(G) \geq c^n/p(n)$ for some polynomial p. Since

$$f_n(G) \leq f_{n-1}(G_\alpha) \leq n f_n(G)$$

for any transitive group G (as G_α-orbits on $(n-1)$-sets correspond to G-orbits on pairs (α, Δ), where Δ is an n-set and $\alpha \in \Delta$), we see that G is in $\mathcal{E}(c)$ if and only if G_α is in $\mathcal{E}(c)$. Now Macpherson shows that, for some $c > 1$, if G is primitive and is not in $\mathcal{E}(c)$, then G is highly set-transitive. In fact, he shows that it is n-set-transitive by induction on n.

The base case $n = 1$ of the induction is vacuous, since any primitive group is transitive.

The first two inductive steps are done separately, and involve the bulk of the work. Suppose that G is not 2-set transitive. Then G preserves a set of 2-sets, which we can take to be the edge set of a graph on the vertex set Ω. First it is shown that the neighbour sets of any two vertices in the graph have infinite symmetric difference. This allows many non-isomorphic subgraphs to be constructed. In fact, using Ramsey's Theorem, Macpherson shows that finite trees can be 'encoded' into orbits of G on finite sets, in such a way that

- the sets in the orbit corresponding to an n-vertex tree have at most $c_1 n$ points, for some constant c_1;
- at most polynomially many trees are encoded into the same orbit.

Now, using the fact that there are exponentially many different trees, elementary bookkeeping shows that G is in $\mathcal{E}(c)$ for some value of c, which can be calculated.

Next Macpherson shows that G is 3-set-transitive, by a similar but even more involved argument, where the graph of the preceding argument is replaced by a 3-uniform hypergraph (whose 'hyperedges' form a G-orbit on 3-sets).

From now on the work is easier, because of two facts.

(a) A group which is 3-set transitive but not 2-transitive preserves a linear order (see Theorem 5.19). It can be shown that, if such a group is not highly homogeneous, then it lies in the class $\mathcal{E}(2)$.

(b) There is a specific group B with the following property: Let G be a permutation group of countable degree which is 3-set-transitive and 2-transitive but not 2-primitive (that is, the point stabiliser is imprimitive). Then either

- G preserves the betweenness relation on $\mathbb{Q}$ (that is, its elements preserve or reverse the order), or
- G is a subgroup of B.

Moreover, B is in $\mathcal{E}(c)$, where $c = 2.483\ldots$ (see Cameron [32, IV]). If G preserves a betweenness relation then, applying (a) to its normal subgroup of index 2 which preserves the order, either G is highly set-transitive or G is in $\mathcal{E}(2)$.

So let G be n-set-transitive for $n \geq 3$, and G not in $\mathcal{E}(c)$ for some c (take the smallest value required in any of the above arguments). Then by (a) and (b) above, we may assume that G is 2-primitive. So G_α is primitive and not in $\mathcal{E}(c)$. By the inductive hypothesis, G is n-set-transitive, whence G is $(n+1)$-set-transitive, and the inductive step is complete.

Remark. The group B is the point stabiliser in the group associated with boron trees in the preceding section. It can be regarded as the automorphism group of the Fraïssé limit of the following class of relational structures. The point set of a structure is the set of leaves of a binary tree; the relation $R(x; y, z)$ holds if the paths from the root to y and z diverge from one another later than they diverge from the path from the root to x (see Figure 5.5). In the Fraïssé limit, the relation $R(x; y, z)$ holds if y and z are in the same block of imprimitivity for the stabiliser of x.

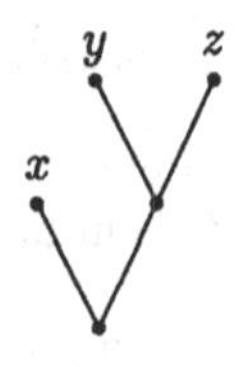

Figure 5.5: Ternary relation from binary tree

For general groups, the following holds. (The first part requires a technical condition, namely, that G is the automorphism group of a homogeneous relational structure in a finite language.) The first part is due to Pouzet [147], the second to Macpherson [128].

Theorem 5.23 *(a) Either $(f_n(G))$ grows like n^k for some $k \in \mathbb{N}$ (in the sense that $an^k \leq f_n(G) \leq bn^k$ for all n, where a and b are constants and $a > 0$), or $(f_n(G))$ grows faster than any polynomial in n.*

(b) In the latter case, $f_n(G) \geq \exp(n^{1/2-\epsilon})$ for all $n \geq n_0(\epsilon)$.

Polynomial growth occurs in imprimitive groups. For example, $f_n(S \mathrm{Wr} S_k)$ is the number of partitions of n into at most k parts, which is asymptotically $n^k/((k-1)!k!)$ for fixed k.

There appears to be a spectrum of possible growth rates between polynomial and exponential, though no restrictions have been proved. If $f_n(G) \sim n^k$, then $f_n(G \text{ Wr } S)$ is roughly $\exp(n^{(k+1)/(k+2)})$. Also, the iterated wreath product G of more than two infinite symmetric groups has the property that $(f_n(G))$ grows faster than any fractional exponential function of the above form, but slower than any exponential function.

Macpherson [130] has some results on growth rates faster than exponential. These turn out to be related to model-theoretic properties such as *stability* and the *strict order property* that cannot be discussed here.

5.12 On complementation and switching

I end this chapter with a theorem about finite graphs, for which the only known proof uses results about infinite permutation groups. This theorem also highlights the importance of the operation of *switching* of graphs, defined below.

Let $\mathcal{F}$ be a finite collection of finite graphs. Given any graph G, we define the *hypergraph* $\mathcal{F}(G)$ as follows: its vertex set is the same as that of G; its *hyperedges* are all the sets U of vertices having the property that the induced subgraph on U is isomorphic to a member of $\mathcal{F}$. (A hypergraph is a structure like a graph except that its 'hyperedges' can have any finite cardinality, not just 2.)

Example 1. If $\mathcal{F}$ consists of the graph having two vertices and one edge, then $\mathcal{F}(G)$ is the same as G. If $\mathcal{F}$ consists of the graph having two vertices and no edges, then $\mathcal{F}(G)$ is the complement of G.

Example 2. If $\mathcal{F}$ consists of all n-vertex graphs, then $\mathcal{F}(G)$ is the complete n-uniform hypergraph on the vertex set of G.

Example 3. Let G_1 and G_2 be graphs on the vertex set X, and let Y be a subset of X. We say that G_2 is obtained by *switching* G_1 with respect to Y if

- the edges of G_1 and G_2 within Y coincide;
- the edges of G_1 and G_2 outside Y coincide;
- for $y \in Y$ and $z \in X \setminus Y$, $\{y, z\}$ is an edge of G_1 if and only if it is not an edge of G_2.

In other words, the edges between Y and its complement are 'switched' into non-edges and *vice versa.* Switching successively with respect to two subsets is the same as switching with respect to their symmetric difference; so switching is an equivalence relation on the set of all graphs on the vertex set X. Its equivalence classes are called *switching classes.* The operation of switching was introduced by Jaap Seidel [163].

Theorem 5.24 *Let $\mathcal{F}$ consist of the complete graph on three vertices together with the graph with three vertices and one edge.*

(a) $\mathcal{F}(G_1) = \mathcal{F}(G_2)$ if and only if G_1 and G_2 are equivalent under switching.

(b) A 3-uniform hypergraph H is of the form $\mathcal{F}(G)$ for some graph G if and only if every 4-set of vertices contains an even number of hyperedges of H.

Proof. None of this is deep: I will prove the reverse implication in (b). Suppose that H satisfies the hypotheses. Select a vertex x, and define a graph G by the rule that $y \sim z$ if and only if $\{x, y, z\}$ is a hyperedge in H. (Note that x is an isolated vertex.) Now H and $\mathcal{F}(G)$ coincide on all triples containing x; and the parity condition implies that $\{u, v, w\}$ is a hyperedge if and only if an odd number of $\{x, u, v\}$, $\{x, u, w\}$, $\{x, v, w\}$ are hyperedges, that is, if and only if an odd number of $\{u, v\}$, $\{u, w\}$, $\{v, w\}$ are edges of G, as required.

A hypergraph satisfying the parity condition of Theorem 5.24(b) was called a *two-graph* by Graham Higman, who introduced the concept. The theorem shows that there is a natural bijection between two-graphs and switching classes of graphs. So the two concepts are equivalent.

Two-graphs have important connections with permutation groups. For example, many of the known finite 2-transitive groups are groups of automorphisms of two-graphs. There is a reason for this, hinted at in the next result. Let H and K be subgroups of G, with $K < H$. We say that K

is *strongly closed* in H if $K^g \cap H \leq K$ for all $g \in G$. This is a necessary condition for the existence of a normal subgroup N of G with $N \cap H = K$. Sometimes it is sufficient. For example, Wielandt proved a 'transfer theorem' asserting, in particular, that if G is a 3-transitive permutation group and G_α has a strongly closed subgroup K of index 2, then G has a normal subgroup N of index 2 with $N_\alpha = K$. If we weaken the assumption of 3-transitivity to 2-transitivity, then either the same conclusion holds, or G acts on a non-trivial two-graph. This shows, for example, that the 2-transitive groups $\mathrm{PSL}(2,q)$ (for $q \equiv 1 \pmod 4$) or $\mathrm{PSU}(3,q)$ (for q odd) act on two-graphs. See Hale and Shult [86], Wielandt [188].

See Seidel [164] for a survey of two-graphs and switching.

Example 4. In any graph, the *chromatic number* (the number of colours required for a proper colouring) is at least as large as the *clique number* (the size of the largest complete subgraph), since vertices of a complete subgraph must get different colours in any proper colouring. Claude Berge [16] defined a *perfect graph* to be one which has the property that the clique number and chromatic number of any induced subgraph are equal. Perfect graphs form an important class of graphs; for example, many algorithmic problems which are hard for general graphs are tractable for perfect graphs. Berge made two conjectures about their structure, the so-called *strong* and *weak perfect graph conjectures*. The WPGC asserts that the complement of a perfect graph is perfect. The SPGC asserts that a graph is perfect if and only if neither it nor its complement contains an induced cycle of odd length greater than 3. The SPGC implies the WPGC, since its conditions are self-complementary.

The WPGC was proved by László Lóvasz [121] (and is now referred to as the *Perfect Graph Theorem*). Later, Bruce Reed [154] proved a *Semi-Strong Perfect Graph Theorem*, which implies the WPGC and is implied by the SPGC. It states that, if $\mathcal{F}(G_1) = \mathcal{F}(G_2)$ and G_1 is perfect, then G_2 is perfect, where $\mathcal{F}$ consists of the path with four vertices and three edges. (Since this graph is self-complementary, we have $\mathcal{F}(G) = \mathcal{F}(\overline{G})$, so the WPGC follows from Reed's Theorem.)

Another consequence of Reed's Theorem is that a graph containing no induced subgraphs of length 3 (an *N-free graph*) is perfect. There are exponentially many such graphs. However, it is a consequence of the main theorem of this section that, for almost all finite graphs G_1, $\mathcal{F}(G_1) = \mathcal{F}(G_2)$ implies that $G_2 = G_1$ or $G_2 = \overline{G_1}$, so Reed's Theorem is no stronger than Lovász's for a 'typical' graph.

In order to state the next theorem, we define five equivalence relations $E_1, \ldots, E_5$ on graphs on a given set of vertices, as follows:

- $(G_1, G_2) \in E_1$ if and only if $G_1 = G_2$.

- $(G_1, G_2) \in E_2$ if and only if $G_1 = G_2$ or $G_1 = \overline{G_2}$.
- $(G_1, G_2) \in E_3$ if and only if G_1 and G_2 are switching-equivalent.
- $(G_1, G_2) \in E_4$ if and only if G_1 is switching-equivalent to G_2 or to $\overline{G_2}$.
- $(G_1, G_2) \in E_5$ for all pairs of graphs with the same vertex set.

We recall that *almost all finite graphs have property* P if the proportion of labelled n-vertex graphs having P tends to 1 as $n \to \infty$. Now the theorem of Cameron and Martins [44] is as follows:

Theorem 5.25 *Let $\mathcal{F}$ be a finite set of finite graphs. Then there is an index $i \in \{1, \ldots, 5\}$ such that, for almost all finite graphs G_1, we have $\mathcal{F}(G_1) = \mathcal{F}(G_2)$ if and only if $(G_1, G_2) \in E_i$.*

Before proving the theorem, we reconsider the examples. In Examples 1, 2, 3, we have $i = 1$, $i = 5$, and $i = 3$ respectively, and in fact the theorem holds with 'almost all' replaced by 'all'. In Example 4, we have $i = 2$; almost all graphs are equivalent (in Reed's sense) only to themselves and their complements.

Proof. We require the following important theorem of Simon Thomas [174], for which the proof is too long to give here.

Theorem 5.26 *Let R be the countable random graph. There are just five closed subgroups of* $\mathrm{Sym}(R)$ *containing* $\mathrm{Aut}(R)$*, namely:* $\mathrm{Aut}(R)$*; the group of automorphisms and anti-automorphisms of R; the group of switching automorphisms of R; the group of switching automorphisms and switching anti-automorphisms of R; and* $\mathrm{Sym}(R)$.

Here an *anti-automorphism* is an isomorphism from R to its complement; a *switching automorphism* is an isomorphism from R to a graph equivalent to it under switching; and the definition of a *switching anti-automorphism* can be imagined. Note the parallel with Corollary 5.18.

We also need the theorem of Glebskii *et al.* (Theorem 5.6) on first-order sentences true in R, and the footnote to the theorem of Engeler *et al.* (Theorem 5.11) on countably categorical structures.

Let $\mathcal{F}$ be given. Consider the group $G = \mathrm{Aut}(\mathcal{F}(R))$. It is a closed subgroup of $\mathrm{Sym}(R)$ containing $\mathrm{Aut}(R)$, and so is one of the five described in Thomas' theorem. We assume that $G = \mathrm{Aut}(R)$; the other cases are similar but a bit more complicated in detail. Note that G is the automorphism group of a hypergraph, which can be represented as a structure in a suitable first-order language (which we call the *hypergraph language*) having relations for the various sizes of hyperedges.

In this case, G has two orbits on ordered pairs of distinct vertices of R, namely edges and non-edges. By Theorem 5.11, there is a first-order formula $\phi(x,y)$ in two variables (in the hypergraph language) which holds precisely when x and y are joined in R.

Now any formula of the hypergraph language can be 'translated' into a formula of the graph language, replacing each hyperedge relation by the disjunction of statements describing all the possible graph structures of graphs in $\mathcal{F}$ of the appropriate cardinality. Let $\tilde{\phi}(x,y)$ be the 'translation', in this sense, of $\phi(x,y)$. Then by assumption, the sentence

$$(\forall x)(\forall y)((x \sim y) \leftrightarrow \tilde{\phi}(x,y))$$

holds in R.

By Theorem 5.6, this sentence holds in almost all finite graphs.

But, in any graph G in which it holds, it tells us exactly how to reconstruct the graph from the hypergraph: x and y are joined in G if and only if $\phi(x,y)$ is true in the hypergraph $\mathcal{F}(G)$. This case of the proof is complete, since this says that for almost all graphs G, if $\mathcal{F}(G) = \mathcal{F}(G')$, then $G = G'$, that is, $(G,G') \in E_1$.

5.13 Appendix: Cycle index

The cycle index of a finite permutation group was originally developed by Redfield, Pólya, de Bruijn, and others in connection with combinatorial enumeration. In this section I give an outline.

Let Ω be a finite set. We assume that associated with each point of Ω there is an object or 'figure', taken from a set F. For example, Ω may be the vertex or edge set of a graph, and F a set of colours applied to the vertices or edges. The set F is not required to be finite; but each figure has a non-negative integer weight, and we require that there are only a finite number of figures of any given weight. The *figure-counting series* is the formal power series

$$A(t) = \sum_{i \geq 0} a_i t^i,$$

where a_i is the number of figures of weight i.

The weight of a function $f : \Omega \to F$ is now defined to be

$$w(f) = \sum_{\alpha \in \Omega} w(f(\alpha)).$$

There are only a finite number, say b_i, of functions of weight i; and the *function-counting series* is

$$B(t) = \sum_{i \geq 0} b_i t^i.$$

Now $B(t) = A(t)^n$, where $n = |\Omega|$. This is easily seen when $|\Omega| = 2$, since there are $\sum_{j=0}^{i} a_j a_{i-j}$ choices of figures with weights summing to i, and this is the coefficient of t^i in $A(t)^2$. The general case is proved similarly, or by induction.

This observation is a special case of the Cycle Index Theorem, which we now state.

Let G be a permutation group on Ω. We are interested in counting the orbits of G on the set of functions from Ω to F. Accordingly, we re-define b_i to be the number of G-orbits on functions of weight i, and let $B(t) = \sum_{i\geq 0} b_i t^i$ be the *function-counting series* for the permutation group G.

Theorem 5.27 (Cycle Index Theorem)

$$B(t) = Z(G; s_i \leftarrow A(t^i)).$$

The cycle index $Z(G)$ is obtained by averaging over G the cycle indices $z(g) = s_1^{c_1(g)} \cdots s_n^{c_n(g)}$ of its elements, where $c_i(g)$ is the number of i-cycles of the permutation g. So the theorem follows from the fact that the number of functions of weight i fixed by the permutation g is the coefficient of t^i in $z(g; s_i \leftarrow t^i)$. This is proved by an extension of the earlier argument. If g is a single i-cycle, then a g-invariant function is constant, and its weight is i times the weight of the figure in its range; so the function-counting series for functions fixed by an i-cycle is $A(t^i)$. Now, by much the same argument as before, we take the product over all the cycles of an arbitrary permutation.

Since the cycle index of the trivial group is s_1^n, our earlier assertion follows.

Example. In how many different ways, up to rotations, can the faces of a cube be coloured with m different colours?

Taking G to be the rotation group of the cube, acting on the faces, we find that the cycle index is

$$Z(G) = \tfrac{1}{24}(s_1^6 + 6s_1^2 s_4 + 3s_1^2 s_2^2 + 8s_3^2 + 6s_2^3).$$

If we take each colour to be a figure of weight 0, then the figure-counting series is simply m, and the number of orbits is

$$\tfrac{1}{24}(m^6 + 3m^4 + 12m^3 + 8m^2).$$

Example. We can count orbits of a permutation group G on n-sets, as follows. Take two figures, with weights 0 and 1. Then any function can be regarded as the characteristic function of a subset of Ω, and its weight is the cardinality of the subset. The figure-counting series is $1 + t$, and so the function-counting series is

$$f_G(t) = \sum_{i\geq 0} f_i(G) t^i = Z(G; s_i \leftarrow 1 + t^i).$$

We note that the cycle index $Z(G)$ is the only polynomial in indeterminates $s_1,\ldots,s_n$ for which the Cycle Index Theorem holds. In fact, it suffices that the theorem holds for all possible figure-counting series with at least one figure of weight 0. From this, we can deduce that

$$\tilde{Z}(G) = Z(G; s_i \leftarrow 1 + s_i)$$

by showing that the polynomial $\tilde{Z}(G; s_i \leftarrow s_i - 1)$ also satisfies the Cycle Index Theorem for such functions. Let F be a set of figures with at least one figure of weight 0. Let $*$ be a distinguished figure of weight 0, and let $F' = F \setminus \{*\}$, with figure-counting series $a(t) - 1$, where $a(t)$ is the figure-counting series for F. Now we count orbits of G on functions of weight i as follows:

- First, choose an orbit of G on subsets of Ω. Let Δ be a representative set in the orbit, and let the function map every point outside Δ to $*$.
- Now let H be the group induced on Δ by its setwise stabiliser, and choose an orbit of H on functions from Δ to F'.

The number of choices of an orbit on functions of weight i in the second step is the coefficient of t^i in $Z(H; s_i \leftarrow a(t^i) - 1)$, by the Cycle Index Theorem. Summing over orbit representatives Δ, and using the fact that $\sum_\Delta Z(H) = \tilde{Z}(G)$, we see that the total number of G-orbits on functions of weight i is the coefficient of t^i in $\tilde{Z}(G; s_i \leftarrow a(t^i) - 1)$. So this is equal to $Z(G; s_i \leftarrow a(t^i))$ for all such figure-counting series $a(t)$. It follows from our remark that $Z(G) = \tilde{Z}(G; s_i \leftarrow s_i - 1)$, as required.

We conclude this appendix with the calculation of the modified cycle index of the infinite symmetric group S. This group has a single orbit on n-sets for every n. So

$$\begin{aligned}
\tilde{Z}(S) &= \sum_{n\geq 0} Z(S_n) \\
&= \sum_{n\geq 0} \sum_{n=a_1+2a_2+\cdots} \frac{n!}{1^{a_1}a_1!\,2^{a_2}a_2!\,\cdots} s_1^{a_1} s_2^{a_2} \cdots \\
&= \sum_{a_1,a_2,\ldots}^{*} \prod_i \frac{1}{a_i!}\left(\frac{s_i}{i}\right)^{a_i} \\
&= \prod_i \sum_a \frac{1}{a!}\left(\frac{s_i}{i}\right)^{a} \\
&= \prod_i \exp\left(\frac{s_i}{i}\right) \\
&= \exp\left(\sum_i \frac{s_i}{i}\right).
\end{aligned}$$

In the third line, $\sum^*$ denotes the sum over all sequences $(a_1, a_2, \ldots)$ such that all but finitely many terms are zero. The step from this line to the next is the 'infinite distributive law'.

For more on cycle index, see Harary and Palmer [88], or Goulden and Jackson [84].

5.14 Exercises

In these exercises, S denotes the symmetric group of countable degree, A the group of order-preserving permutations of $\mathbb{Q}$, and C the group of permutations of the set of complex roots of unity preserving the cyclic order (a transitive extension of A).

5.1. A permutation of the vertex set of a graph is said to be *almost an automorphism* if it maps edges to edges and non-edges to non-edges with only finitely many exceptions. Prove that the group of almost automorphisms of the random graph R is highly transitive.

5.2. Prove that R has the *pigeonhole property*: if its vertex set is partitioned into two parts, then the induced subgraph on one of the parts is isomorphic to the original graph. Show further that the only countable graphs with the pigeonhole property are the complete and null graphs and the random graph.

5.3. Show that each of the following graphs is isomorphic to R:

- (for set theorists) Take a countable model of the Zermelo–Fraenkel axioms for set theory. Form a graph by joining x to y if and only if either $x \in y$ or $y \in x$.

- (for number theorists) Take the set of primes congruent to 1 mod 4. Join p to q if and only if p is a quadratic residue mod q (or q is a quadratic residue mod p – it is the same thing, by the Law of Quadratic Reciprocity).

5.4. Let X be a subset of the set $\mathbb{N}$ of positive integers, and define a graph $\Gamma(X)$ as follows: the vertex set is $\mathbb{Z}$, and x and y are joined if and only if $|x - y| \in X$.

(a) Show that $\Gamma(X)$ has an automorphism t_X which permutes its vertices in a single cycle.

(b) Show that, if X is chosen at random by including natural numbers independently with probability $1/2$, then $\Gamma(X)$ is isomorphic to R with probability 1. Deduce that R admits cyclic automorphisms.

(c) Suppose that $\Gamma(X) \cong \Gamma(X')$. Prove that t_X and $t_{X'}$ are conjugate in $\mathrm{Aut}(\Gamma(X))$ if and only if $X = X'$. Deduce that R admits $2^{\aleph_0}$ non-conjugate cyclic automorphisms.

5.5. Prove that a permutation group G on Ω is oligomorphic if and only if, for any $\alpha_1, \ldots, \alpha_n \in \Omega$, the pointwise stabiliser of $\alpha_1, \ldots, \alpha_n$ has only finitely many orbits on Ω.

5.6. Let G be a countable $\aleph_0$-categorical group. (According to Theorem 5.10, this means that $\mathrm{Aut}(G)$ acts oligomorphically on G.)

(a) Prove that G is a locally finite group of finite exponent.

(b) Prove that G has only finitely many *characteristic* subgroups (subgroups fixed by all automorphisms).

5.7. Prove that the stabilisers of all n-tuples of points form a basis of open neighbourhoods of the identity for the topology on the symmetric group of countable degree. Show that these subgroups are closed, and deduce that the topology is totally disconnected.

5.8. Prove that a subgroup of $\mathrm{Sym}(\Omega)$ is discrete if and only if there exist $\alpha_1, \ldots, \alpha_n$ such that the pointwise stabiliser of these points is the identity.

5.9. Prove

- $F_{G\times H}(t) = F_G(t)F_H(t)$;
- $f_{G\times H}(t) = f_G(t)f_H(t)$;
- $F_{G\,\mathrm{Wr}\,H}(t) = F_H(F_G(t) - 1)$.

Show that $f_{G\,\mathrm{Wr}\,H}(t)$ is not determined by $f_G(t)$ and $f_H(t)$.

5.10. Calculate the modified cycle index $\tilde{Z}(C)$, and the generating functions $f_C(t)$ and $F_C(t)$, where C is the group of permutations preserving the cyclic order on the complex roots of unity. Hence show that $F_n(C \,\mathrm{Wr}\, S) = F_n(A)$ for all n. Use the cycle decomposition of a permutation to give an alternative proof of this fact.

5.11. Using the result of the preceding exercise, and the fact that (since C is highly homogeneous) we have $\tilde{Z}(C; t, t^2, \ldots) = f_C(t) = 1/(1-t)$, prove that

$$\prod_{k\geq 1}(1 - t^k)^{-\phi(k)/k} = \exp(t/(1-t)),$$

where $\phi(k)$ is Euler's function.

5.12. Prove that

$$f_{G \mathrm{Wr} S}(t) = \prod_{k \geq 1} (1 - t^k)^{-f_k(G)}.$$

5.13. Prove that $f_n(S \,\mathrm{Wr}\, A)$ is the number of unlabelled *preorders* (reflexive and transitive relations satisfying trichotomy) on n points.

5.14. Let G be a finite permutation group. Prove that $\tilde{Z}(G)$ is obtained from $Z(G)$ by substituting $s_i + 1$ for s_i for all i.

5.15. Show that the kth component P_k of the Parker vector of a finite permutation group G is given by the expression

$$P_k = k \left(\frac{\partial \tilde{Z}(G)}{\partial s_k} \right) \quad (s_i \leftarrow 0 \text{ for all } i).$$

What happens for infinite oligomorphic groups?

5.16. Prove that, if a permutation group is 4-set-transitive and 2-transitive but not 3-transitive, then it preserves either a circular order, or the betweenness relation derived from a linear order. You may assume, or prove if you can, the following results:

(a) A ternary relation on a set of at least four points is a circular order if and only if its restriction to every 4-element subset is a circular order;

(b) A ternary relation on a set of at least four points is the betweenness relation derived from a linear order if and only if its restriction to every 4-element subset is the betweenness relation derived from a linear order.

5.17. Show that the five cases in Theorem 5.25 are determined by $\mathcal{F}$ as follows:

- case E_1 if $\mathcal{F}$ is not closed under complementation or switching;
- case E_2 if $\mathcal{F}$ is closed under complementation but not under switching;
- case E_3 if $\mathcal{F}$ is closed under switching but not under complementation;
- case E_4 if $\mathcal{F}$ is closed under complementation and switching but, for some m, contains some but not all m-vertex graphs;
- case E_5 if, for every m, $\mathcal{F}$ contains all or no m-vertex graphs.

5.18. (a) Show that there is a graph B with vertex set $X \cup Y$, where X and Y are disjoint countable sets, with the property that a *random bipartite graph* on $X \cup Y$ (obtained by choosing edges between X and Y independently with probability 1/2) is almost surely isomorphic to B. Show further that B is characterised by the properties that

(i) for all finite disjoint subsets U, V of X, there is a vertex $z \in Y$ joined to all vertices in U and to none in V;

(ii) same with X and Y interchanged.

(b) Show that the subgroup of $\mathrm{Aut}(B)$ fixing the sets X and Y is highly transitive on each of them.

(c) Let Γ be any bipartite graph on $X \cup Y$, with bipartite blocks X and Y. Suppose that the subgroup of $\mathrm{Aut}(\Gamma)$ fixing X and Y is highly transitive on each of them. Show that one of the following occurs:

(i) Γ or its bipartite complement is null;

(ii) Γ or its bipartite complement is a matching;

(iii) Γ is isomorphic to B.

[The *bipartite complement* of Γ has edge set consisting of all edges between X and Y which are not in Γ. A *matching* is a bipartite graph in which each vertex of X is joined to one vertex of Y and *vice versa.*]

5.19. This exercise gives Rado's construction [153] of the random $(k+1)$-uniform hypergraph.

(a) Let k be given. Show that there is a bijection ϕ from the set of k-subsets of $\mathbb{N}$ to the set $\mathbb{N}$, as follows: if $x_1 < \cdots < x_k$, then set

$$\phi(\{x_1, \ldots, x_k\}) = \binom{x_1}{1} + \cdots + \binom{x_k}{k}.$$

(b) Consider the set of all $(k+1)$-subsets $\{x_1, \ldots, x_k, z\}$ of $\mathbb{N}$ for which $x_1 < \cdots < x_k < z$ and the $\phi(\{x_1, \ldots, x_k\})$ digit in the base-2 expansion of z is 1. Show that these sets form the random $(k+1)$-uniform hypergraph on the set $\mathbb{N}$.

5.20. In how many ways can the edges of a cube be coloured with m colours up to rotations of the cube?

5.21. Prove Theorem 5.13.

5.22. Let S be the symmetric group on an infinite set.

(a) Show that $f_n(S \mathrm{\,Wr\,} S_2) = f_n(S_2 \mathrm{\,Wr\,} S) = \lfloor (n+1)/2 \rfloor$.

(b) Show that $f_n(S \mathrm{\,Wr\,} S_k) = f_n(S_k \mathrm{\,Wr\,} S)$ and show that this number is asymptotically $n^k/(k!(k-1)!)$.

(c) If $G = S_k \mathrm{\,Wr\,} S$, show that

$$\sum_{n \geq 0} f_n(G) t^n = \frac{1}{(1-t)(1-t^2)\cdots(1-t^k)}.$$

5.23. (a) Show that $f_n(C_2 \operatorname{Wr} A)$ is the nth Fibonacci number, where A is the group of order-preserving permutations of $\mathbb{Q}$.

(b) Show that $f_n(A \operatorname{Wr} A \operatorname{Wr} \cdots \operatorname{Wr} A) = k^n$, where there are k factors in the iterated wreath product.

CHAPTER 6

Miscellanea

6.1 Finitary permutation groups

A permutation of an infinite set Ω is *finitary* if it moves only finitely many points. The finitary permutations of Ω form a subgroup of $\mathrm{Sym}(\Omega)$, the *finitary symmetric group* $\mathrm{FS}(\Omega)$, which contains the *alternating group* $\mathrm{Alt}(\Omega)$. (If a permutation moves only finitely many points, then we can assign a parity to it in the usual way, independent of the number of fixed points.) A permutation group is called *finitary* if it is contained in the finitary symmetric group.

The finitary symmetric and alternating groups are normal subgroups of the symmetric group. In the case when Ω is countable, they are the only proper normal subgroups; and the factor groups in the normal series

$$1 \trianglelefteq \mathrm{Alt}(\Omega) \trianglelefteq \mathrm{FS}(\Omega) \trianglelefteq \mathrm{Sym}(\Omega)$$

are simple groups, of orders $\aleph_0$, 2 and $2^{\aleph_0}$ respectively. For larger cardinalities, there are further normal subgroups, the *bounded symmetric groups* $\mathrm{BS}_\lambda(\Omega)$, where $\mathrm{BS}_\lambda(\Omega)$ consists of all permutations moving fewer than λ points, where λ is an infinite cardinal not exceeding $|\Omega|$.

The classification of primitive finitary permutation groups was completely settled by Wielandt [184], in a result sometimes called the *Jordan–Wielandt Theorem.* This name derives from a theorem of Jordan, according to which there is a function f such that, if G is a finite primitive permutation group of degree $n \geq f(m)$ which contains a non-identity element moving m points, then G contains the alternating group A_n. Wielandt gave a short proof which extends to infinite groups. The proof of the Jordan–Wielandt theorem given here is due to John Dixon.

Theorem 6.1 *An infinite primitive permutation group which contains a non-identity finitary permutation contains the alternating group.*

Proof. We need three preliminary facts.

Fact 1. A primitive group G containing a 3-cycle contains the alternating group.

Write $\alpha \sim \beta$ if either $\alpha = \beta$ or there is a 3-cycle $(\alpha\ \beta\ \gamma) \in G$. This relation is reflexive and symmetric. A short calculation with permutations on at most five points shows that it is also transitive, and indeed that if $\alpha \sim \beta \sim \gamma$, then $(\alpha\ \beta\ \gamma) \in G$. So $\sim$ is a congruence, and is not equality (by assumption); so all pairs are congruent, and (by the above observation) all 3-cycles lie in G. But the 3-cycles generate the alternating group, so we are done.

Fact 2. If G is transitive and has finite rank, then the union of the finite G_α-orbits is a block for G.

Write $\alpha \sim \beta$ if β lies in a finite G_α-orbit. This time $\sim$ is symmetric and transitive. (For, if $\alpha \sim \beta \sim \gamma$, then $|G_\alpha : G_{\alpha\beta}|$ and $|G_\beta : G_{\beta\gamma}|$ are finite, so $|G_{\alpha\beta} : G_{\alpha\beta\gamma}|$ is finite, so $|G_\alpha : G_{\alpha\gamma}|$ is finite, so $\alpha \sim \gamma$.) Now the set $B(\alpha) = \{\beta : \alpha \sim \beta\}$ is the union of finitely many finite G_α-orbits, so is finite. By transitivity, $\beta \in B(\alpha)$ implies $B(\beta) \subseteq B(\alpha)$. But $|B(\alpha)| = |B(\beta)|$, since an element of G mapping α to β takes $B(\alpha)$ to $B(\beta)$. So these sets are equal, and $\alpha \in B(\beta)$. This means that $\sim$ is symmetric, and so is a congruence.

Fact 3. Let G be a permutation group on Ω with no finite orbits. Let Γ and Δ be finite subsets of Ω. Then there exists $g \in G$ such that $\Gamma g \cap \Delta = \emptyset$.

This is *Neumann's Lemma* (Neumann [138], Lemma 2.3). See the next section for the proof.

Proof of the theorem. Let g be an element of G with non-empty finite support Δ. Take $\alpha \in \Delta$. We claim that every orbit of G_α meets Δ. For, if Γ is one which does not, then Γ is fixed by both G_α and g, and hence by G (since G_α is a maximal subgroup of G), a contradiction.

Thus, G has finite rank, and Fact 2 shows that G_α has no finite orbits except α. By Fact 3, there exists $h \in G_\alpha$ such that $(\Delta \setminus \{\alpha\}) \cap (\Delta \setminus \{\alpha\})h = \emptyset$. In other words, $\Delta \cap \Delta h = \{\alpha\}$. Now Δ and Δh are the supports of g and $h^{-1}gh$ respectively, and a short calculation shows that, if the supports of two permutations meet in just one point, then the commutator of the two permutations is a 3-cycle. Thus G contains a 3-cycle, and so contains the alternating group, by Fact 1.

Remark. Let G be a transitive finitary permutation group. If Δ is a proper block for G, then some element of G maps Δ to a disjoint set; so Δ is finite. If there exists a maximal proper block Δ, then G induces a primitive cofinitary group on its congruence class, necessarily the symmetric or alternating group. On the other hand, if no proper block is maximal, then Ω is the union of an ascending chain of blocks, so Ω is countable. For further details, see Neumann [138, 139], where a structure theory for finitary permutation groups is developed.

6.2 Neumann's Lemma

Neumann's Lemma is the following assertion:

Theorem 6.2 *Let G act on Ω with no finite orbits. Let Γ and Δ be finite subsets of Ω. Then there exists $g \in G$ with $\Gamma g \cap \Delta = \emptyset$.*

Proof. The proof is by induction on $k = |\Gamma|$. The result is trivial for $k = 0$, and clear for $k = 1$: if $\Gamma = \{\gamma\}$, then since the G-orbits are infinite, there is an element of G carrying γ outside Δ. The inductive hypothesis will assert that the theorem is true when $|\Gamma| = k - 1$, for any finite set Δ'.

We may assume that $\Gamma \not\subseteq \Delta$ (translating by an element of G if necessary). Take $\gamma \in \Gamma \setminus \Delta$, and let $\Gamma' = \Gamma \setminus \{\gamma\}$. Then $|\Gamma'| = k - 1$. Let $|\Delta| = m$. Using the induction hypothesis, choose elements $g_1, \ldots, g_{m+1} \in G$ such that

$$\Gamma' g_i \cap (\Delta \cup \Delta g_1 \cup \cdots \cup \Delta g_{i-1}) = \emptyset$$

for $i = 1, \ldots, m + 1$. Now, if $\gamma g_i \notin \Delta$ for some i, then $\Gamma g_i \cap \Delta = \emptyset$, and we are done. So assume that $\gamma g_1, \ldots, \gamma g_{m+1} \in \Delta$. Since $|\Delta| = m$, we must have $\gamma g_i = \gamma g_j$ for some $j < i$. Then $\gamma g_i g_j^{-1} = \gamma \notin \Delta$, and $\Gamma' g_i g_j^{-1} \cap \Delta = \emptyset$ (since $\Gamma' g_i \cap \Delta g_j = \emptyset$). So $\Gamma g \cap \Delta = \emptyset$, where $g = g_i g_j^{-1}$, and we are done.

This proof is due to Birch *et al.* [19]. Another 'combinatorial' proof appears in [34]. However, Neumann's original proof deduced the result from a theorem of his father [137]:

Theorem 6.3 *Let G be a group, $H_1, \ldots, H_k$ subgroups of G, and $x_1, \ldots, x_k$ elements of G. Suppose that $G = H_1 x_1 \cup \cdots \cup H_k x_k$. Then some subgroup H_i has finite index in G.*

The proof of Neumann's Lemma from this theorem runs as follows. Suppose that $\Gamma g \cap \Delta \neq \emptyset$ for all $g \in G$. For each $\gamma \in \Gamma$ and $\delta \in \Delta$, let

$$X_{\gamma\delta} = \{g \in G : \gamma g = \delta\}.$$

Then $G = \bigcup X_{\gamma\delta}$, and each non-empty set $X_{\gamma\delta}$ is a coset of G_γ. By Theorem 6.3, some G_γ has finite index in G, that is, some point γ lies in a finite orbit.

Remark. The deduction of Theorem 6.3 from Theorem 6.2 is equally simple. Let $H_1, \ldots, H_k$ be subgroups of G, and $x_1, \ldots, x_k \in G$ satisfy $G = \bigcup H_i x_i$. Let $\Omega = \bigcup H_i \backslash G$, $\Gamma = \{H_1, \ldots, H_k\}$, $\Delta = \{H_1 x_1, \ldots, H_k x_k\}$. If $g \in G$ satisfies $\Gamma g \cap \Delta = \emptyset$, then $g \notin H_1 x_1 \cup \cdots \cup H_k x_k$. So no such g can exist. By Theorem 6.2, some orbit $H_i \backslash G$ is finite.

6.3 Cofinitary permutation groups

A permutation is *cofinitary* if it fixes only finitely many points. A permutation group G is *cofinitary* if every *non-identity* element of G is cofinitary. Whereas finitary permutation groups trace their lineage to Jordan's Theorem, cofinitary permutation groups descend from such examples as Frobenius groups, *Zassenhaus groups* (2-transitive groups in which the stabiliser of any three points is the identity), etc.

Unlike the situation in the last section, there are many strange primitive cofinitary groups, and no hope of a classification. We will take a trip to the zoo and look at a couple of the creatures. For many more, see the survey [37].

A *Tarski monster* is an infinite group G having the property that all nontrivial proper subgroups have order p, where p is a prime. The existence of such groups was only hypothesised by Tarski. They were constructed (for sufficiently large p) by Ol'šanskii [141]. Further details can be found in his book [142]. In particular, there are Tarski monsters in which all the subgroups of order p are conjugate.

If G is such a group, consider G acting by conjugation on the set of subgroups of order p. Each subgroup of order p fixes itself. As in the proof of Sylow's Theorem in Chapter 1, no such subgroup can normalise another, since if it did then the product would be a group of order p^2. So G is transitive and every non-identity element has exactly one fixed point. This example shows how badly Frobenius' Theorem can fail for infinite groups: a Tarski monster is simple and is generated by any two elements not in the same cyclic subgroup.

A similar but less frightening example is the group $\mathrm{PSU}(2, \mathbb{C})$ of 2×2 unitary matrices modulo scalars. Such a matrix has just two, orthogonal, eigenvectors, unless it is a scalar multiple of the identity. So, in its action on the projective line (the set of 1-dimensional subspaces of $\mathbb{C}^2$), the group is transitive and every non-identity element fixes two points.

Here is an example, due to Kantor, of a highly transitive group. Let $F = \langle f_1, f_2, \ldots \rangle$ be a free group of countable rank, acting regularly on Ω. Make a countable list consisting of all pairs $r_i = (p_i, q_i)$, where p_i and q_i are n-tuples of distinct elements of Ω (for the same n): the list includes all such pairs, for all $n \geq 1$.

For each i, choose a *finitary* permutation m_i mapping the n-tuple $p_i f_i$ to the n-tuple q_i. (This is possible since the finitary symmetric group is highly transitive.) Now let $G = \langle f_1 m_1, f_2 m_2, \ldots \rangle$. We claim that G is the required group.

- *G is highly transitive:* For $p_i f_i m_i = q_i$, so we have included among the generators an element mapping any n-tuple to any other.
- *G is cofinitary:* Let w be a non-trivial word in the generators of G. Since all the m_i belong to the normal subgroup $\mathrm{FS}(\Omega)$ of $\mathrm{Sym}(\Omega)$, we have
$$w(f_1 m_1, f_2 m_2, \ldots) = w(f_1, f_2, \ldots) m$$
for some $m \in \mathrm{FS}(\Omega)$. Now, since F is a free group acting regularly and w a non-trivial word, $w(f_1, f_2, \ldots)$ is fixed-point-free. Then $w(f_1 m_1, f_2 m_2, \ldots)$ is the product of a fixed-point-free and a finitary permutation, and so is cofinitary.

In fact, we see that any non-trivial word in the generators of G is different from the identity; so G is free on the given generators.

In Section 1.12, we proved the theorem of Tits (Theorem 1.11) asserting that certain cofinitary permutation groups don't exist, namely (infinite) sharply 4-transitive groups (and hence also sharply k-transitive groups for $k \geq 4$). However, if we relax the conditions slightly, then the groups do exist:

Theorem 6.4 *(a) For any $k < l$, there is an infinite permutation group which is k-transitive but not $(k+1)$-transitive, and in which the stabiliser of any l points is trivial, but some $(l-1)$-point stabiliser is non-trivial.*

(b) For any k, there is a sharply k-set-transitive group of order-preserving permutations of $\mathbb{Q}$.

Remark. In (a), such groups with $k = l$ would be sharply k-transitive.

Part (b) is due to Macpherson [131], who gave a bare-hands construction. It answers an old problem in the theory of measurement in mathematical psychology, as follows.

The process of measurement is a method of assigning numbers to the elements of some set of objects in the real world. We assume that the objects are ordered, and that the numbers should be assigned in such a way as to reflect this order. In most cases, there is nothing absolute about the number assigned to any object: a different choice of units, for example, could give different numerical measures. Accordingly, a *scale of measurement* is defined by a group G of order-preserving permutations of the number system: it consists of all images of a given assignment of numbers to objects under post-multiplication by the group G. So each group defines a scale of measurement.

We say that a scale of measurement has *scale type* k if a measurement is completely determined by the values it assigns to any k objects, which can be arbitrary subject only to preserving the order. This translates into the assertion that the group G is sharply k-transitive.

The classical scales of measurement are:

- *nominal scales* such as counting, where the number assigned to an object is unique: type 0.
- *ratio scales*, determined up to a factor (the standard example is length, whose numerical value depends on the chosen unit): type 1.
- *interval scales*, of type 2: an example is temperature, before the concept of absolute zero was discovered, when two points (such as the freezing and boiling point of water) were needed to fix the scale.
- *ordinal scales*, where the numbers have no significance apart from defining the order: type ∞.

A theorem of Alper [3] implies that these are the only possible scale types if the real numbers are used for measurement. However, Macpherson's result says that scales of every possible type k are possible if the rationals are used. This raises the philosophical question: why, if in practice any measurement ever made results in a rational number, have the strange scales constructed by Macpherson never been observed?

6.4 Theorems of Blichfeldt and Maillet

The following old theorem of Blichfeldt [22] in 1904 was forgotten for many years.

Theorem 6.5 *Let G be a permutation group of degree n. Let*

$$L = \{\text{fix}(g) : g \in G, g \neq 1\} = \{l_1, \ldots, l_r\},$$

and let $f_L(x) = (x - l_1) \cdots (x - l_r)$. Then $|G|$ divides $f_L(n)$.

Proof. The permutation character π is given by $\pi(g) = \text{fix}(g)$. Now let $\theta(g) = f_L(\pi(g))$. Then θ is a *generalised character*, or *virtual character*, of G (a linear combination of characters with integer coefficients). Also,

$$\theta(g) = \begin{cases} f_L(n) & \text{if } g = 1, \\ 0 & \text{otherwise.} \end{cases}$$

So the multiplicity of the principal character of G in θ is given by

$$\langle \theta, 1 \rangle = \frac{1}{|G|} f_L(n).$$

Of course, this multiplicity is an integer!

Remarks. (a) In fact θ is a multiple of the *regular character* ρ_G (the permutation character of the regular action of G), which is given by

$$\rho_G(g) = \begin{cases} |G| & \text{if } g = 1, \\ 0 & \text{otherwise.} \end{cases}$$

This gives another proof of Theorem 6.5.

(b) The conclusion remains true for any set L of integers which contains the fixed-point numbers of all non-identity elements of G.

The theorem was rediscovered by Ito and Kiyota [97], who investigated groups attaining the bound (that is, groups with $f_L(n) = |G|$), under the name *sharp permutation groups.* The name was suggested by the fact that a sharply k-transitive permutation group is sharp, with $L = \{0, 1, 2, \ldots, k-1\}$.

Problem. Determine the sharp permutation groups (or at least those with $|L| > 1$).

Can Theorem 6.5 be extended to the infinite? Clearly not in the form given! But an alternative, equivalent form looks more promising.

Theorem 6.6 *With the hypotheses of Blichfeldt's Theorem, let*

$$f_L(x) = a_r x^r + \cdots + a_1 x + a_0,$$

and let m_i be the number of orbits of G on Ω^i for $0 \le i \le r$. Then

$$\sum_{i=0}^{r} a_i m_i \ge 0.$$

Proof. We have $\theta = \sum_{i=0}^{r} a_i \pi^i$, and so

$$\langle \theta, 1 \rangle = \sum_{i=0}^{r} a_i \langle \pi^i, 1 \rangle = \sum_{i=0}^{r} a_i m_i.$$

Note that

$$\sum_{i=0}^{r} a_i m_i = f_L(n)/|G|.$$

The form of Theorem 6.6 makes sense for infinite groups, at least for those with $L = \{\text{fix}(g) : g \neq 1\}$ finite (these groups are *cofinitary*!) and with finitely many orbits on Ω^r, where $r = |L|$. But unfortunately it is false. In the last section, we saw transitive groups with $L = \{1\}$ (Tarski monsters) or $L = \{2\}$ (unitary groups); these have $m_0 = m_1 = 1$ and $f_L(x) = x - 1$ or $x - 2$ respectively; so $\sum a_i m_i = 0$ or -1.

However, there is an infinite version of the theorem. The key turns out to be an even older theorem due to Maillet [132], though we only need part of Maillet's insight. Maillet proved Blichfeldt's Theorem without character theory (which was not available to him in 1895!), but with the set L replaced by a larger set

$$L^* = \{\text{fix}(H) : H \leq G, H \neq \{1\}\},$$

the set of all fixed point numbers of *subgroups* of G. Note that L^* contains L, but the maximal elements of these sets are the same. For example, if $G = \text{PGL}(2, \mathbb{C})$, the *Möbius group* of linear fractional transformations of the extended complex plane, then $L = \{1, 2\}$, $L^* = \{0, 1, 2\}$.

In fact, Maillet needed less than this, namely just the fixed point numbers of subgroups of prime power order. Blichfeldt incorrectly cited Maillet's theorem by taking the fixed point numbers of elements rather than subgroups, and regarded his own theorem as a new proof rather than a strengthening.

The following infinite version is proved in [37].

Theorem 6.7 *Let G be a cofinitary permutation group on Ω. Let*

$$L^* = \{\text{fix}(H) : H \leq G, H \neq \{1\}\},$$

and assume that L^* is finite, say $L^* = \{l_1, \ldots, l_r\}$. Let

$$f_{L^*}(x) = (x - l_1) \cdots (x - l_r) = a_r x^r + \cdots + a_1 x + a_0.$$

Suppose that the number m_i of G-orbits on Ω^i is finite for $i \leq r$. Then G is an IBIS group, and the number of orbits of G on irredundant bases is $\sum_{i=0}^{r} a_i m_i$; so, in particular, this number is a positive integer.

Example. For $G = \text{PGL}(2, \mathbb{C})$, we have $L^* = \{0, 1, 2\}$, $f_{L^*}(x) = x^3 - 3x^2 + 2x$, $m_0 = m_1 = 1$, $m_2 = 2$, $m_3 = 5$; so

$$\sum_{i=0}^{3} a_i m_i = 5 - 6 + 2 - 0 = 1,$$

in agreement with the fact that G is sharply 3-transitive.

Geometric groups, which we considered in Section 4.14, are precisely those groups which satisfy the hypotheses of the above theorem with $\sum a_i m_i = 1$. As we noted there, the finite geometric groups with $r > 1$ have been completely determined by Tracey Maund (using CFSG), whereas the determination of finite sharp groups is still open. Here is an intermediate problem, motivated by Maillet's Theorem. Let $L^\dagger$ be the set of fixed point numbers of non-trivial subgroups of the finite permutation group G which have prime power order. Then, replacing L^* by $L^\dagger$ in the above theorem, it is again true that $\sum a_i m_i > 0$ (or that $|G|$ divides $f_{L^\dagger}(n)$). Now which groups have $\sum a_i m_i = 1$ (or $|G| = f_{L^\dagger}(n)$)?

6.5 Cycle-closed permutation groups

This problem arose in the work of Lenart and Ray on Hopf algebras. Let G be a finite permutation group on Ω. Let $C(G)$ be the group generated by all the cycles of the elements of G. We say that G is *cycle-closed* if $C(G) = G$. The original question was, which groups are cycle-closed? To make things a bit more interesting, define $C_0(G) = G$ and $C_{n+1}(G) = C(C_n(G))$ for $n \geq 0$.

Theorem 6.8 *(a) The finite permutation group G is cycle-closed if and only if G is a direct product of symmetric groups and cyclic groups of prime order.*

(b) For any finite permutation group G, $C_3(G)$ is cycle-closed.

(c) There exist groups G for which $C_2(G)$ is not cycle-closed. Such a group, if transitive, is a p-group for some odd prime p.

An example under (c) is the cyclic group G of order p^2 acting regularly, where p is an odd prime. We have $C_1(G) = C_p \operatorname{Wr} C_p$, $C_2(G) = A_{p^2}$, and $C_3(G) = S_{p^2}$. A complete determination of the groups G with $C_2(G) \neq C_3(G)$ is not known.

There are infinite generalisations of the notion of cycle-closure; in fact, several possible generalisations, since in general $G \not\leq C(G)$ for an infinite permutation group G. These are considered in [39]. In particular, the group $C(G)$, where G is the infinite cyclic group acting regularly, is a particularly interesting highly transitive permutation group.

6.6 Fixed-point-free elements

It is a trivial consequence of the Orbit-Counting Lemma that:

> If G is a finite transitive permutation group with degree greater than 1, then G contains a fixed-point-free element.

For the average number of fixed points is 1, and the identity fixes more than one point. An equivalent group-theoretic formulation is:

> If H is a proper subgroup of the finite group G, then there is a conjugacy class in G which is disjoint from H.

Fein, Kantor and Schacher [72] proved the following extension.

Theorem 6.9 [CFSG] *If G is a finite transitive permutation group with degree greater than 1, then G contains a fixed-point-free element of prime power order.*

Equivalently, in the second formulation, the elements of the conjugacy class disjoint from H can be taken to have prime power order.

There are two remarkable features of this theorem. First, it is used to prove a number-theoretic result, the main theorem of Fein, Kantor and Schacher:

Theorem 6.10 [CFSG] *The relative Brauer group of any finite extension of global fields is infinite.*

Moreover, this fact implies the theorem as stated.

Digression. Here are the definitions necessary for the Fein–Kantor–Schacher Theorem.

- A *global field* is defined to be a finite extension of either $\mathbb{Q}$ or $F(t)$, where F is a finite field and t is transcendental over F.
- The Brauer group $B(K)$ of a field K is defined as follows: first consider the semigroup formed by the isomorphism types of central simple algebras over K (the simple K-algebras with centre K), with operation given by $(A, B) \mapsto A \otimes_K B$. It contains a sub-semigroup consisting of the matrix algebras $M_n(K)$ for $n \geq 1$; the quotient is the *Brauer group* $B(K)$.
- Let L be an extension of K. There is a map $B(K) \to B(L)$, given by $A \mapsto A \otimes_K L$; its kernel (the set of equivalence classes of K-algebras which 'split' over L) is the *relative Brauer group* $B(L/K)$.

The second feature is that the result requires CFSG, and no small amount of work on the finite simple groups. We outline the proof. Take a minimal counterexample G to the theorem: that is, G has minimal degree and, subject to that, minimal order, among transitive groups with no fixed-point-free elements of prime power order. Then

- *G is primitive*: otherwise the group induced on a congruence class would be a smaller counterexample.

- *G is simple*: for a proper normal subgroup would be transitive and hence a smaller counterexample.

Now we invoke CFSG to conclude that G is a known simple group, and examine the simple groups in turn. To give the flavour, here is a sketch of the argument in the case $G = \mathrm{PSL}(d,q)$ for $d > 2$. A lot of detail has been left out here, and what is left might be confusing! Let $H = G_\alpha$, a maximal subgroup of G. By assumption, H meets every conjugacy class of elements of prime power order in G.

First, assume that $(d,q) \neq (6,2)$ (this case must be treated separately). We use *Zsigmondy's Theorem*, an essential tool in many arguments about classical groups. A *primitive divisor* of $a^N - 1$ is a prime divisor of $a^N - 1$ which does not divide $a^i - 1$ for $1 \leq i \leq N-1$.

Theorem 6.11 *A primitive divisor of $a^N - 1$ exists for all $a \geq 2$, $N \geq 2$, except for the cases $N = 2$, $a = 2^k - 1$, and $N = 6$, $a = 2$.*

A proof can be found in Lüneburg [122].

It follows that, if r is a primitive divisor of $q^d - 1$, then G, and hence H, contains elements of order r; so H is irreducible. (For the eigenvalues of such an element are rth roots of unity, whose minimal polynomial over $\mathrm{GF}(q)$ has degree d.)

A *transvection* is an element of G induced by a linear map of the form $x \mapsto x + (x\phi)a$ of $V = \mathrm{GF}(q)^d$, where ϕ is an element of the dual space V^* and a an element of V which satisfy $a\phi = 0$. This element has order p, the characteristic of the field (the prime divisor of q). So H contains a transvection.

Let H^* be the subgroup of H generated by the transvections in H. Then Frattini's Lemma shows that $H = H^* N_H(P)$, where P is a Sylow p-subgroup of H. From this and the irreducibility of H, we conclude that H^* is also irreducible. We appeal to the classification of irreducible groups generated by transvections to identify H^*, and hence H (which is the normaliser of H^* in G). In each case, we find that G is not a counterexample after all. For example, H might be the symplectic group $\mathrm{PSp}(d,q)$. Now, for most choices of d and q, there is a primitive divisor of $q^{d-1} - 1$, which divides $|G|$ and not $|H|$ (looking at the order formulae for these groups, which you can find in Table 7.1).

There are many hard open problems connected with this result. For example, which prime? We would expect that, if one prime power dominates the degree of G, there would be a fixed-point-free element of order a power of this prime. More formally:

Conjecture. (Generalised Isbell Conjecture) There is a function $f(p,k)$ such that, if G is a transitive permutation group of degree $n = p^a k$ with $p \not| k$ and $a \geq f(p,k)$, then G contains a fixed-point-free element of p-power order.

This was conjectured by Isbell in 1959, in connection with game theory, in the case $p = 2$. It is still open nearly forty years later. It has been settled for primitive groups, but this does not help, since we cannot do induction in this case. Another approach is to try to prove the following stronger conjecture:

Conjecture. There is a function $g(p,k)$ such that if P is a p-group with k orbits each of size at least $p^{g(p,k)}$, then P contains a fixed-point-free element.

The earlier conjecture would follow on taking P to be a Sylow p-subgroup of G: its orbits would all have size at least the p-part of n, and so the number of them would be at most k. Only very small cases of this conjecture have been proved: if P has at most p non-trivial orbits, or at most $3p/2$ orbits of size at least p^2, then it does contain a fixed-point-free element (the second assertion is a theorem of Aron Bereczky [15]).

In Theorem 6.9, the element cannot be chosen to have prime order. Consider, for example, the affine group

$$G = \{x \mapsto ax + b : a, b \in \mathrm{GF}(9), a \neq 0\},$$

acting on the set of 12 *lines* of the affine plane of order 3. The elements of prime order in G have order 2 (reflections, conjugate to $x \mapsto -x$, which fix all four lines through the origin) or 3 (translations, $x \mapsto x + b$, which fix three parallel lines).

However, a conjecture about this was made independently by Dragan Marušič and Mikhail Klin. A permutation group G on Ω is called *2-closed* if every permutation of Ω which preserves all the G-orbits on Ω^2 lies in G. For example, the full automorphism group of a graph (possibly directed) is 2-closed.

Conjecture. A transitive 2-closed permutation group contains a fixed-point-free element of prime order.

6.7 The Orbit-Counting Lemma revisited

Nigel Boston drew my attention to a neat reformulation of the Orbit-Counting Lemma.

Theorem 6.12 *Let G be a finite permutation group. Let $X(t)$ be the probability generating function for the number of fixed points of a random element of G (that is, $X(t) = \sum p_k t^k$, where p_k is the fraction of elements of G having exactly k fixed points), and let $F(t) = \sum F_k t^k / k!$, where F_k is the number of*

G-orbits on k-tuples of distinct elements. Then

$$X(t) = F(t-1).$$

Corollary 6.13 *With the above notation, the proportion of fixed-point-free elements in G is $F(-1)$.*

Proof. If an element g of G fixes k points, then it fixes $k(k-1)\cdots(k-i+1)$ i-tuples of distinct points. So the Orbit-Counting Lemma gives

$$F_i = \frac{1}{|G|}\sum_{k=0}^{n} p_k|G|k(k-1)\cdots(k-i+1).$$

Multiplying by $t^i/i!$ and summing over i gives

$$\begin{aligned} F(t) &= \sum_{i=0}^{n} F_i t^i/i! \\ &= \sum_{i=0}^{n}\sum_{k=i}^{n} p_k \binom{k}{i} t^i \\ &= \sum_{k=0}^{n} p_k \sum_{i=0}^{k} \binom{k}{i} t^i \\ &= \sum_{k=0}^{n} p_k(1+t)^k = X(t+1), \end{aligned}$$

from which the result follows.

Example. For the group $G = S_n$, we have $F_k = 1$ for $0 \le k \le n$, so $F(t)$ is the series for e^t truncated to degree n. Thus, $X(t)$ is approximately e^{t-1}, from which we see that, if k is small relative to n, the probability that a random element has k fixed points is roughly $1/\mathrm{e}k!$. In other words, the distribution of the number of fixed points of a random element of S_n tends to the Poisson with parameter 1 as $n \to \infty$. In particular, the proportion of *derangements* (fixed-point-free elements) tends to $1/\mathrm{e}$, a well-known result.

Arratia and Tavaré [4] proved a generalisation of this result. (Their theorem is stronger than given here.)

Theorem 6.14 *For fixed k, the joint distribution of the numbers of 1-cycles, 2-cycles, ..., k-cycles of a random element of S_n tends to that of independent Poisson variables with parameters 1, 1/2, ..., $1/k$ as $n \to \infty$.*

This means that the probability that a random element has a_i cycles of length i for $i = 1, \ldots, k$ tends to

$$\prod_{i=1}^{k} \frac{1}{a_i!\, i^{a_i}} \mathrm{e}^{-1/i}$$

as $n \to \infty$.

Proof. It is possible to derive this result from a more general version of Parker's Lemma in a similar way to the derivation of the distribution of the number of fixed points from the Orbit-Counting Lemma. We give a different, and briefer, argument, based on cycle index techniques. Let the required probability be $p(a_1,\ldots,a_k)$. Then

$$\begin{aligned}
\sum_{a_1,\ldots,a_k} p(a_1,\ldots,a_k)s_1^{a_1}\cdots s_k^{a_k} &= \frac{1}{n!}\sum_{g\in S_n} s_1^{c_1(g)}\cdots s_k^{c_k(g)} \\
&= Z(S_n; s_i \leftarrow 1 \text{ for } i>k) \\
&\to \tilde{Z}(S; s_i \leftarrow s_i - 1 \text{ for } i \le k, \\
&\qquad s_i \leftarrow 0 \text{ for } i>k) \\
&= \exp\left((s_1-1)+\frac{s_2-1}{2}+\cdots+\frac{s_k-1}{k}\right)
\end{aligned}$$

as $n\to\infty$. (We use the facts that

$$Z(S_n) = \tilde{Z}(S_n; s_i \leftarrow s_i - 1) = \sum_{m=0}^{n} Z(S_m; s_i \leftarrow s_i - 1),$$

and $\tilde{Z}(S) = \sum_{m\ge 0} Z(S_m)$.) From this, the required convergence of the coefficients can be read off. (The expression on the right is precisely the probability generating function for k independent Poisson variables with parameters $1, 1/2, \ldots, 1/k$.)

Remark. As we saw in the last chapter, the power series $F(t) = F_G(t)$ is defined for any oligomorphic permutation group G. In many interesting cases, either this series converges at $t=-1$, or it can be summed there by a standard method. For example, if S, A, C denote the infinite symmetric group, the group of order-preserving permutations of $\mathbb{Q}$, and the group of permutations preserving the cyclic order of the roots of unity, we have:

$$F_S(t) = \sum_{n\ge 0}\frac{t^n}{n!} = e^t, \qquad F_S(-1) = 1/e = 0.367\,879\,441,$$

$$F_A(t) = \sum_{n\ge 0} t^n = 1/(1-t), \qquad F_A(-1) = \tfrac{1}{2} = 0.5,$$

$$F_C(t) = 1 + \sum_{n\ge 0}\frac{t^n}{n} = 1-\log(1-t), \qquad F_C(-1) = 1-\log 2 = 0.306\,852\,819.$$

It is not at all clear what these strange numerical invariants of oligomorphic groups mean!

6.8 Jordan groups

Let G be a permutation group on Ω. A subset Δ of Ω containing more than one point is called a *Jordan set* if there is a subgroup of G which fixes $\Omega \setminus \Delta$ pointwise and acts transitively on Δ. (We exclude singleton sets Δ, which would trivially satisfy this condition.) Since a Jordan set is contained within a single orbit of G, there is no great loss in assuming that G is transitive. Then Ω is a Jordan set; any other Jordan set is called *proper*. (This terminology is not standard: see below.)

In this section, we let $G(\Delta)$ denote the pointwise stabiliser of the complement of the Jordan set Δ. It acts transitively on Δ.

If G is k-transitive, then any set Δ for which $|\Omega \setminus \Delta| < k$ is a Jordan set. We call G a *Jordan group* if it is transitive and has a Jordan set not of this type. (The usual terminology is that a Jordan set Δ is improper if G is $(|\Omega \setminus \Delta| + 1)$-transitive. I prefer to use the term 'proper' for a proper subset which is a Jordan set.)

All the finite primitive Jordan groups have been determined. (See Theorem 6.17 below for a partial result on this.) The first proof was that of Kantor [106]. From Jordan's Theorem below, we know that such groups are 2-transitive; we can now invoke CFSG and check through the known list of 2-transitive groups. This is what Kantor did. Subsequently, a proof not using CFSG was found by Evans [68]. This was a considerable achievement, given the effort that had been expended on this question before CFSG. Evans' techniques are geometric; we will develop the theory far enough to see where the geometry comes from. Note that geometric groups are rich in Jordan sets: the pointwise stabiliser of any set, if non-trivial, is transitive on the points it moves. We will see that there is a geometry (or matroid) associated with a finite primitive Jordan group which generalises the geometry we have met in a geometric group.

If two Jordan sets Δ_1 and Δ_2 intersect, then their union is a Jordan set: the group $\langle G(\Delta_1), G(\Delta_2)\rangle$ acts transitively on it and fixes its complement pointwise. It follows that, if Δ_1 is a maximal proper Jordan set, then one of the following must occur:

- $\Delta_1 \cap \Delta_2 = \emptyset$;
- $\Delta_2 \subseteq \Delta_1$;
- $\Delta_1 \cup \Delta_2 = \Omega$.

Also, if B is a block and Δ a Jordan set with non-empty intersection, then one of B and Δ contains the other. For, if $\alpha \in B \cap \Delta$, $\beta \in B \setminus \Delta$, and $\delta \in \Delta \setminus B$, then a permutation mapping α to δ cannot fix β.

The key result in the theory is *Jordan's Theorem*:

Theorem 6.15 *A finite primitive group having a proper Jordan set is 2-transitive.*

Proof. We begin with a simple observation.

> If G is primitive on Ω, and Δ is a proper subset of Ω containing more than one point, then for all $\alpha, \beta \in \Omega$, there exists $g \in G$ with $\alpha \notin \Delta g$, $\beta \in \Delta g$.

For, if no such g exists, then the relation $\sim$, defined by $\alpha \sim \beta$ if every translate of Δ containing α contains β, would be a non-trivial congruence.

Now let $|\Omega| = n$, and let the cardinality of a maximal proper Jordan set Δ be k. By our observation on primitivity, for any α, the translates of Δ not containing α cover $\Omega \setminus \{\alpha\}$; and since no two of these sets have union Ω, they partition $\Omega \setminus \{\alpha\}$, so that k divides $n - 1$.

Since G is transitive, any point α is outside the same number $(n-1)/k$ of translates of Δ. So the total number of translates is $n(n-1)/k(n-k)$. Since k divides $n-1$, both k and $n-k$ are coprime to n; so $k(n-k)$ divides $n-1$. This is only possible if $k = 1$ or $k = n-1$, and the first violates the definition of a Jordan set. So $k = n-1$ and G is 2-transitive.

This result enables us to make a stronger statement about the relation between Jordan sets and blocks in an arbitrary group: *a maximal Jordan set is either a block or the complement of one.* For let Δ be a maximal Jordan set, B a block meeting Δ. If $B = \Delta$, the assertion is true. If $\Delta \subset B$, then for all $g \in G$, we have $\Delta \cup \Delta g \neq \Omega$; so distinct translates of Δ are disjoint, and Δ is a block. Finally, if $B \subset \Delta$, then the set of blocks contained in Δ is a Jordan set for the primitive group induced on the set of translates of B by G, and so Δ contains all but one block.

Furthermore, we see that G permutes transitively its maximal blocks.

Now we are in the position to do induction: if $\Delta_1 \subset \Delta_2$ are Jordan sets, then Δ_2 is a Jordan set for $G(\Delta_1)$. In this situation, we say that Δ_2 *covers* Δ_1 if there is no Jordan set lying strictly between Δ_1 and Δ_2. Then $G(\Delta_2)$ permutes transitively the Jordan sets covered by Δ_2. So G permutes transitively the maximal chains

$$\Delta_1 \subset \cdots \subset \Delta_r = \Omega$$

of Jordan sets in which each term covers its predecessor.

One more observation is necessary to construct the geometry. If G is primitive then, whenever Δ_1 and Δ_2 are Jordan sets such that Δ_2 covers Δ_1, then Δ_1 is the complement of a block for $G(\Delta_2)$, rather than a block. This can be proved in several ways. Most easily, it follows from a theorem of Marggraff asserting that, in a primitive group other than the symmetric or alternating group, the cardinality of a Jordan set is at least $|\Omega|/2$. (See Wielandt [186], for example.)

Theorem 6.16 *Let G be a finite primitive permutation group on Ω, and suppose that there exists a Jordan set for G. Then the complements of the Jordan sets, together with Ω, form the subspace lattice of a matroid, admitting G as a group of automorphisms, which acts transitively on the bases and on the maximal flags of subspaces.*

Proof. Let $\mathcal{F}$ consist of Ω and the complements of Jordan sets. To show that $\mathcal{F}$ is a meet-semilattice, it suffices to show that it is closed under pointwise intersection. So take $F_1, F_2 \in \mathcal{F}$. If $F_1 \cap F_2 = \emptyset$, there is nothing to prove. If $|F_1 \cap F_2| \geq 1$, then $\Omega \setminus F_1$ and $\Omega \setminus F_2$ are subsets of $\Omega \setminus \{\alpha\}$ of size at least $|\Omega|/2$ (by Marggraff's Theorem), and so they intersect; so

$$\Omega \setminus (F_1 \cap F_2) = (\Omega \setminus F_1) \cup (\Omega \setminus F_2)$$

is a Jordan set, and $F_1 \cap F_2 \in \mathcal{F}$. The fact that $\mathcal{F}$ is ranked follows from the transitivity of G on maximal chains, noted above. The final condition in Theorem 4.24 is clear from the fact that, if $\rho(F) = r$, then the subspaces of rank $r+1$ containing F have the form $F \cup B$, where B is a block of imprimitivity in a (unique) congruence for $G(\Omega \setminus F)$.

In view of the above result, we speak of the *rank* of a Jordan group (or, more generally, a group admitting a Jordan set), meaning the rank of the associated matroid (which is equal to the length of the longest chain of subspaces).

Projective and affine groups provide examples of primitive Jordan groups. The associated matroids are just the projective and affine spaces.

Theorem 6.17 [CFSG] *All finite primitive groups admitting proper Jordan sets are known. In particular, the only such groups of rank at least 6 are:*

(a) symmetric and alternating groups;

(b) subgroups of $\mathrm{P\Gamma L}(n,q)$ *containing* $\mathrm{PSL}(n,q)$;

(c) subgroups of $\mathrm{A\Gamma L}(n,q)$ *containing* $\mathrm{ASL}(n,q)$.

Note that this includes the assertion that the only finite 6-transitive groups are symmetric and alternating groups.

What about infinite Jordan groups? We begin with two small observations.

Theorem 6.18 *(a) An infinite primitive group containing a finite Jordan set contains the alternating group.*

(b) An infinite primitive but not highly transitive group in which every finite set lies in the complement of a cofinite Jordan set is a projective or affine group over a finite field.

Proof. Part (a) is immediate from the Jordan–Wielandt Theorem. Part (b) follows from Theorem 6.17 using the inductive properties of Jordan sets. This is sometimes known as the *Cherlin–Mills–Zil'ber Theorem*: it was proved specifically for an application in model theory (the classification of $\aleph_0$-categorical, ω-stable theories by Cherlin, Harrington and Lachlan [50]). We give another application of it in the next section.

Jordan's Theorem fails in the infinite case. The simplest counterexample is the group A of order-preserving permutations of $\mathbb{Q}$, in which any open interval is a Jordan set. This group is 2-set-transitive but not 2-transitive.

In a major recent result, Adeleke and Macpherson [1] have given a complete classification of primitive Jordan groups which are not highly transitive, in terms of the relations they preserve (orders, treelike relations, etc.) The relations are discussed in detail by Adeleke and Neumann [2].

Highly transitive Jordan groups, other than bounded symmetric groups (or alternating groups), do exist. Examples include the homeomorphism groups of various topological spaces such as manifolds, where (for example) a connected component of the complement of a submanifold of codimension 1 is a Jordan set.

A good account of infinite Jordan groups appears in Bhattacharjee *et al.* [17].

6.9 Orbits on moieties

In this section, all permutation groups will have countable degree. If Ω is a countable set, we define a *moiety* of Ω to be an infinite subset of Ω with infinite complement, that is, a set which is neither finite nor cofinite. We now present the most important results about orbits on the set of moieties.

Theorem 6.19 *The number of orbits of G on moieties is at least as great as the number of orbits on n-sets, for any natural number n.*

This result extends the monotonicity result of Section 5.9 into the infinite. The second (Ramsey theory) proof applies almost unchanged. If the n-subsets of an infinite set are coloured with r colours, all of which are used, then (after possibly re-ordering the colours $c_1, \ldots, c_r$) there are moieties $Y_1, \ldots, Y_r$ such that Y_i contains a set of colour c_i but none of colour c_j for $j > i$. Thus, if the colour classes are orbits (or unions of orbits) of G, then $Y_1, \ldots, Y_r$ lie in different G-orbits.

Theorem 6.20 *Let G be a primitive permutation group on the countable set Ω which has no finite or countable orbits on moieties. Then either G is highly transitive, or it is a projective or affine group over a finite field.*

This result is due to Neumann [140].

Proof. The hypothesis that G has no finite or countable orbits on moieties is inherited by subgroups of finite or countable index in G (in particular, by the pointwise stabiliser of any finite set). Moreover, a group with this property must have only finitely many finite orbits and a single infinite orbit on Ω, else it would fix a moiety. Let Γ be a finite set, and Γ' the union of the finite orbits of its stabiliser. Then the complement of Γ' is a Jordan set disjoint from Γ. Now the result follows from Theorem 6.18.

Theorem 6.21 *Let G be a primitive permutation group on the countable set Ω which has fewer than $2^{\aleph_0}$ orbits on moieties. Then G is highly transitive.*

This theorem was proved by Macpherson [127] at the same time, and using the same methods, as his result on exponential growth of the number of orbits on n-sets. A completely different proof was later found by Evans [69].

6.10 Exercises

6.1. In this exercise, we quantify the Jordan–Wielandt theorem in order to prove Jordan's Theorem.

(a) Show that there is a function $g(r,s)$ such that, if the finite primitive group G has rank r and G_α has a non-trivial orbit of size s, then the degree of G is at most $g(r,s)$. [*Hint:* by Higman's Theorem, the orbital graph has valency s and is connected with diameter at most r.]

(b) Show that if G is transitive on Ω and $|\Omega| > |\Gamma| \cdot |\Delta|$, then the conclusion of Neumann's Lemma holds.

(c) More generally, show that, if the G-orbits are $\Omega_1, \ldots, \Omega_r$, if $|\Omega_i| = n_i$, $|\Omega_i \cap \Gamma| = c_i$ and $|\Omega_i \cap \Delta| = d_i$ for $i = 1, \ldots, r$, and $\sum_{i=1}^{r} c_i d_i / n_i < 1$, then the conclusion of Neumann's Lemma holds.

Hence prove *Jordan's Theorem*: there is a function f such that, if G is a finite primitive permutation group of degree $n \geq f(m)$ which contains a non-identity element moving m points, then G contains the alternating group A_n.

6.2. Let G be a group of finitary permutations. Prove that G acts faithfully on any infinite orbit. Prove that, if G is transitive, then any non-trivial congruence has finite classes, and that the stabiliser of all the classes is contained in the direct product of a family of finite groups.

[*Note:* The *direct product* of a family $(G_i : i \in I)$ of groups is the subgroup of the cartesian product consisting of all functions $f : I \to \bigcup_{i \in I} G_i$ with $f(i) \in G_i$ for all i and $f(i) = 1$ for all but finitely many values of i.]

6.3. Prove Theorem 6.8. Prove also that, if G is a cyclic group of odd prime power order p^n $(n > 1)$ acting regularly, then $C_2(G) \neq C_3(G)$.

6.4. Let G be a transitive p-group of degree greater than 1. Prove that at least $1 + ((p-1)/p)|G|$ elements of G are fixed-point-free.

[*Note:* This was improved to $\lfloor (p/(p+1))|G| \rfloor$ by Cameron, Kovács, Newman and Praeger [42]; this is best possible.]

6.5. Show, using the preceding exercise, that if G is a p-group with at most p orbits, all non-trivial, then G contains a fixed-point-free element.

6.6. (a) Complete the proof of the Fein–Kantor–Schacher Theorem 6.9 for the alternating groups. You may find the following facts useful:

- if $n \geq 8$, there exists a prime p satisfying $n/2 < p < n-2$ (*Bertrand's postulate*);[1]
- a primitive group of degree n containing a p-cycle for $p < n-2$ is alternating or symmetric (this can be proved using Jordan's Theorem).

(b) Complete the proof of the Fein–Kantor–Schacher theorem for a finite simple group (or a class of such groups) of your choice.

6.7. Let α_n be the proportion of fixed-point-free elements in a Sylow 2-subgroup of S_{2^n}. Prove that

$$\alpha_{n+1} = \tfrac{1}{2}(1 + \alpha_n^2).$$

Deduce that $\alpha_n \to 1$ as $n \to \infty$.

6.8. Prove the following extension of Parker's Lemma. Let $k_1, k_2, \ldots, k_r$ be pairwise coprime. Then the number of orbits of a permutation group G on elements of S_n which are the product of cycles of length $k_1, \ldots, k_r$ and which occur as part of the cycle decomposition of an element of G is

$$\frac{1}{|G|} \sum_{g \in G} k_1 c_{k_1}(g) \cdots k_r c_{k_r}(g).$$

6.9. Prove that, for fixed k, the proportion of fixed-point-free elements in the permutation group S_n on k-element subsets tends to a limit ϵ_k as $n \to \infty$. Show that $\epsilon_2 = 2e^{3/2} = 0.446\,260\,320\ \ldots$, and find ϵ_3. [*Hint:* Theorem 6.14.]

[1] You can find a short proof of this by Robin Chapman at the URL

`http://www.maths.ex.ac.uk/~rjc/etc/bertrand.dvi`

6.10. Let G be a permutation group on a countable set Ω.

(a) Show that, if G has fewer than $2^{\aleph_0}$ orbits on moieties, then G is oligomorphic.

(b) Show that, if G has only finitely many orbits on moieties, and G is transitive, then G is highly transitive.

6.11. Show that any base for G and the support of any non-identity element of G have non-empty intersection.

Hence show that, if G is transitive on Ω, with $|\Omega| = n$, and G has *minimal degree* m (the smallest cardinality of the support of a non-identity element of G) and minimal base size b, then $mb \geq n$.

6.12. Use this result and Babai's bound on the base size of a uniprimitive permutation group G of degree n (Section 4.10) to give a lower bound, not depending on CFSG, for the minimal degree of such a group.

Remark. Stronger results on minimal degree, using CFSG, are given by Liebeck and Saxl [118].

6.13. We say that the permutation group G has *type* L if L is the set of numbers of fixed points of non-identity elements of G.

(a) Suppose that G has type $\{m\}$. Prove that G has at least $m+1$ orbits, with equality if and only if it is sharp.

(b) If G is a sharp group of type $\{1\}$, prove that G is geometric (with one fixed point and one regular orbit).

(c) Let G be a finite group of rotations of Euclidean 3-space, acting on the set of points where its rotation axes meet the unit sphere. Prove that G is sharp with type $\{2\}$.

(d) Prove that any sharp group of type $\{2\}$ with all orbit lengths greater than 2 is one of those described in (c).

6.14. Some questions related to Maillet's Theorem.

(a) Prove that, if G is a finite permutation group of degree n, and L^* the set of fixed point numbers of non-trivial subgroups of G, then $|G|$ divides $\prod_{l \in L^*}(n-l)$. [*Hint:* Induction on $|L^*|$.]

(b) Prove *Maillet's Theorem*: if G is a finite permutation group of degree n, and $L^\dagger$ the set of fixed point numbers of non-trivial subgroups of G of prime power order, then $|G|$ divides $\prod_{l \in L^\dagger}(n-l)$. [*Hint:* Show that it suffices to prove the theorem for groups of prime power order.]

(c) Prove Theorem 6.7. (You may use [37] as a crib.)

6.15. An *independence algebra* is an algebra A (in the sense of universal algebra, that is, a set with a family of operations) satisfying the following conditions:

(a) The subalgebras of A satisfy the *exchange property*, that is, if $z \in \langle Y \cup \{y\} \rangle$ and $z \notin \langle Y \rangle$, then $y \in \langle Y \cup \{z\} \rangle$. [This implies that the subalgebras are the closed sets of a matroid, so that, in particular, all *bases* (minimal generating sets) have the same cardinality.]

(b) Any map from a base of A into A can be extended to an endomorphism of A.

Prove that the automorphism group of A is a geometric group.

Remark. Using Maund's classification of the finite geometric groups (see Section 4.14), Cameron and Szabó [48] have determined all finite independence algebras up to equivalence, where two algebras are equivalent if there is a bijection between them preserving the subalgebras and endomorphisms.

6.16. Let G be a group and C a set on which G acts 'on the left' (so that $(g_1g_2)(c) = g_1(g_2(c))$ for all $g_1, g_2 \in G$, $c \in C$). Let $A = G \cup C$, and define operations on A as follows:

- for each $c \in C$, a nullary operator ν_c with value c;
- for each $g \in G$, a unary operator λ_g given by

$$\begin{aligned} \lambda_g(h) &= gh \quad \text{for } h \in G; \\ \lambda_g(c) &= g(c) \quad \text{for } c \in C. \end{aligned}$$

Prove that A is an independence algebra. Prove also that any independence algebra with a base of size 1 is equivalent to one of this form (possibly with $C = \emptyset$).

CHAPTER 7

Tables

In this chapter, some information about the finite simple groups and the finite 2-transitive groups is tabulated.

The first two tables list the non-abelian finite simple groups, excluding the alternating groups. Table 7.1 gives the groups of Lie type, and Table 7.2 the sporadic groups.

The following tables list the finite 2-transitive groups, subdivided according to Burnside's Theorem. Table 7.3 gives the 2-transitive groups with abelian regular normal subgroups, and Table 7.4 those whose minimal normal subgroup is simple.

Further information about the finite simple groups can be found in the ATLAS *of Finite Groups* [54], or in the on-line Atlas of Group Representations, at

`http://www.mat.bham.ac.uk/atlas/`

7.1 Simple groups of Lie type

This table lists the finite simple groups of Lie type.

There is one family of *Chevalley groups* for each simple Lie algebra over $\mathbb{C}$. These algebras are classified by the *Dynkin diagrams* A_n, B_n, C_n, D_n, E_6, E_7, E_8, F_4, G_2 (see Figure 7.1). The Chevalley groups are defined over all fields: the parameter q denotes the order of the finite field. The *twisted groups* are associated with *graph automorphisms* of Dynkin diagrams. The superscript t indicates the order (2 or 3) of the graph automorphism, which must be combined with a field automorphism of the same order. For Dynkin diagrams with only single bonds, the field automorphism must have the same order as the graph automorphism; the *Steinberg groups* ${}^2A_n$, ${}^2D_n$, ${}^3D_4$, ${}^2E_6$ are obtained, and are defined over fields of order q^t. If there is a multiple bond, the characteristic of the field is equal to the number (2 or 3) of bonds, and the square of the field automorphism is equal to the Frobenius map; so the field order is an odd power of the characteristic. This yields the *Suzuki and Ree groups* ${}^2B_2$, ${}^2G_2$ and ${}^2F_4$.

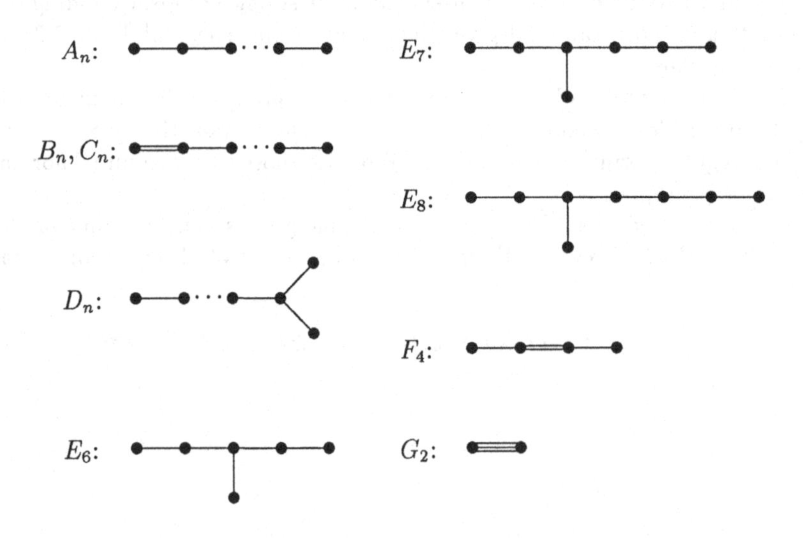

Figure 7.1: Dynkin diagrams

For further information about these groups, including their definition, see Carter [49].

In the table, n is a positive integer, m a non-negative integer, q a prime power. The group occurs as a quotient of a matrix group over GF(q) of order

N by a central subgroup of order d (and so has order N/d).

All groups in the table are simple except for

- $A_1(2) \cong S_3$,
- $A_1(3) \cong A_4$,
- $B_2(2) \cong S_6$,
- $G_2(2)$, which has a subgroup of index 2 isomorphic to ${}^2A_2(3^2)$),
- ${}^2A_2(2^2)$, which is a Frobenius group of order 72,
- ${}^2B_2(2)$, which is a Frobenius group of order 20,
- ${}^2G_2(3)$, which has a subgroup of index 3 isomorphic to $A_1(2^3)$,
- ${}^2F_4(2)$, which has a simple subgroup of index 2, the *Tits group*, not occurring elsewhere in the table.

Other coincidences between groups of Lie type and alternating groups are

- $A_1(2^2) \cong A_1(5) \cong A_5$,
- $A_1(7) \cong A_2(2)$,
- $A_1(3^2) \cong A_6$,
- $A_3(2) \cong A_8$,
- $B_2(3) \cong {}^2A_3(2^2)$.

Note that $C_n(q)$ is defined for $n = 2$ and for q even but is isomorphic to $B_n(q)$ in these cases; similarly $D_3(q)$ and ${}^2D_3(q^2)$ are isomorphic to $A_3(q)$ and ${}^2A_3(q^2)$ respectively.

Another subdivision of the groups of Lie type is into classical and exceptional groups. The *classical groups* are the linear groups and subgroups preserving forms: symplectic groups preserve alternating bilinear forms, orthogonal groups preserve quadratic forms, and unitary groups preserve hermitian forms. In each case, we take the derived group – which is usually but not always the subgroup of matrices having determinant 1 – and factor out the central subgroup of scalar matrices. There are six types, each depending on a dimension and field order: the orthogonal groups must be separated into three types corresponding to different kinds of quadratic forms. The identification is as follows:

- $A_n(q) \cong \mathrm{PSL}(n+1, q)$,

- $B_n(q) \cong \mathrm{P}\Omega(2n+1, q)$,
- $C_n(q) \cong \mathrm{PSp}(2n, q)$,
- $D_n(q) \cong \mathrm{P}\Omega^+(2n, q)$,
- ${}^2A_n(q^2) \cong \mathrm{PSU}(n+1, q)$,
- ${}^2D_n(q^2) \cong \mathrm{P}\Omega^-(2n, q)$.

See Dieudonné [62] or Taylor [171] for information on the classical groups.

The remaining eight families, each depending only on a field order, are the *exceptional groups.* They behave rather like classical groups of bounded dimension.

Other common names are $Sz(q)$ for the Suzuki group ${}^2B_2(q)$, and $R_1(q)$ and $R_2(q)$ for the Ree groups ${}^2G_2(q)$ and ${}^2F_4(q)$ respectively.

The classical groups can be treated geometrically. The group $\mathrm{PSL}(n+1, q)$ acts on the *projective space* $\mathrm{PG}(n, q)$ whose elements are the subspaces of an $(n+1)$-dimensional vector space over $\mathrm{GF}(q)$. The other classical groups act on *polar spaces* whose objects are the subspaces which are *totally isotropic* or *totally singular* (that is, on which the relevant form vanishes identically). These methods have been extended to all the groups of Lie type using Tits' concept of a *building* [176]. See Brown [26], Ronan [155] or Scharlau [159] for details.

	Group	Condition	N	d
Chevalley groups	$A_n(q)$	$n \geq 1$	$q^{n(n+1)/2} \prod_{i=1}^{n} (q^{i+1} - 1)$	$(n+1, q-1)$
	$B_n(q)$	$n \geq 2$	$q^{n^2} \prod_{i=1}^{n} (q^{2i} - 1)$	$(2, q-1)$
	$C_n(q)$	$n \geq 3$, q odd	$q^{n^2} \prod_{i=1}^{n} (q^{2i} - 1)$	$(2, q-1)$
	$D_n(q)$	$n \geq 4$	$q^{n(n-1)}(q^n - 1) \prod_{i=1}^{n-1} (q^{2i} - 1)$	$(4, q^n - 1)$
	$E_6(q)$		$q^{36}(q^{12}-1)(q^9-1)(q^8-1)(q^6-1)(q^5-1)(q^2-1)$	$(3, q-1)$
	$E_7(q)$		$q^{63}(q^{18}-1)(q^{14}-1)(q^{12}-1)(q^{10}-1)(q^8-1)(q^6-1)(q^2-1)$	$(2, q-1)$
	$E_8(q)$		$q^{120}(q^{30}-1)(q^{24}-1)(q^{20}-1)(q^{18}-1)(q^{14}-1)(q^{12}-1)(q^8-1)(q^2-1)$	1
	$F_4(q)$		$q^{24}(q^{12}-1)(q^8-1)(q^6-1)(q^2-1)$	1
	$G_2(q)$		$q^6(q^6-1)(q^2-1)$	1
Steinberg groups	${}^2A_n(q^2)$	$n \geq 2$	$q^{n(n+1)/2} \prod_{i=1}^{n} (q^{i+1} - (-1)^{i+1})$	$(n+1, q+1)$
	${}^2D_n(q^2)$	$n \geq 4$	$q^{n(n-1)}(q^n + 1) \prod_{i=1}^{n-1} (q^{2i} - 1)$	$(4, q^n + 1)$
	${}^2E_6(q^2)$		$q^{36}(q^{12}-1)(q^9+1)(q^8-1)(q^6-1)(q^5+1)(q^2-1)$	$(3, q+1)$
	${}^3D_4(q^3)$		$q^{12}(q^8+q^4+1)(q^6-1)(q^2-1)$	1
Suzuki and Ree groups	${}^2B_2(q)$	$q = 2^{2m+1}$	$q^2(q^2+1)(q-1)$	1
	${}^2G_2(q)$	$q = 3^{2m+1}$	$q^3(q^3+1)(q-1)$	1
	${}^2F_4(q)$	$q = 2^{2m+1}$	$q^{12}(q^6+1)(q^4-1)(q^3+1)(q-1)$	1

Table 7.1: Groups of Lie type

7.2 Sporadic simple groups

This table lists the twenty-six sporadic finite simple groups. Further properties of these groups are in the **ATLAS** *of Finite Groups* [54].

Some of these groups have alternative names: for example,

- $Co_n = .n$ for $n = 1, 2, 3$,
- $Fi_n = M(n)$ for $n = 22, 23, 24$,
- $F_1 = FG = M$ (the *Friendly Giant* or *Monster*),
- $F_2 = BM = B$ (the *Baby Monster*),
- $F_3 = Th$,
- $F_5 = HN$,
- $Mc = McL$,
- $J_2 = HJ$.

Mathieu found his groups in the 1870s; all the others were discovered in the 1960s and 1970s. Mathieu simply wrote down generating permutations; it is not clear how he found them. In the other cases, the process is better documented. Usually, the discovery of evidence for the group was separate from its actual construction (which often involved extensive computation). The discovery often resulted from case analysis related to CFSG, typically information about the centraliser of an involution or the Sylow 2-subgroup (or, in Fischer's case, orders of pairwise products of involutions in a conjugacy class). In many cases, the groups were constructed as automorphism groups of orbital graphs in their smallest permutation representations; in others, as groups of matrices over finite fields or, for the Conway groups, automorphism groups of the *Leech lattice* in 24-dimensional real space. Sometimes (for example, Higman–Sims, McLaughlin, Conway), discovery and construction were simultaneous; in other cases (the Monster, J_4), there was a gap of several years.

Group	Name	Order
M_{11}	Mathieu	$2^4 \cdot 3^2 \cdot 5 \cdot 11$
M_{12}	Mathieu	$2^6 \cdot 3^3 \cdot 5 \cdot 11$
M_{22}	Mathieu	$2^7 \cdot 3^2 \cdot 5 \cdot 7 \cdot 11$
M_{23}	Mathieu	$2^7 \cdot 3^2 \cdot 5 \cdot 7 \cdot 11 \cdot 23$
M_{24}	Mathieu	$2^{10} \cdot 3^3 \cdot 5 \cdot 7 \cdot 11 \cdot 23$
J_1	Janko	$2^3 \cdot 3 \cdot 5 \cdot 7 \cdot 11 \cdot 19$
J_2	Hall-Janko	$2^7 \cdot 3^3 \cdot 5^2 \cdot 7$
J_3	Janko	$2^7 \cdot 3^5 \cdot 5 \cdot 17 \cdot 19$
J_4	Janko	$2^{21} \cdot 3^3 \cdot 5 \cdot 7 \cdot 11^3 \cdot 23 \cdot 29 \cdot 31 \cdot 37 \cdot 43$
Co_1	Conway	$2^{21} \cdot 3^9 \cdot 5^4 \cdot 7^2 \cdot 11 \cdot 13 \cdot 23$
Co_2	Conway	$2^{18} \cdot 3^6 \cdot 5^3 \cdot 7 \cdot 11 \cdot 23$
Co_3	Conway	$2^{10} \cdot 3^7 \cdot 5^3 \cdot 7 \cdot 11 \cdot 23$
Fi_{22}	Fischer	$2^{17} \cdot 3^9 \cdot 5^2 \cdot 7 \cdot 11 \cdot 13$
Fi_{23}	Fischer	$2^{18} \cdot 3^{13} \cdot 5^2 \cdot 7 \cdot 11 \cdot 13 \cdot 17 \cdot 23$
Fi'_{24}	Fischer	$2^{21} \cdot 3^{16} \cdot 5^2 \cdot 7^3 \cdot 11 \cdot 13 \cdot 17 \cdot 23 \cdot 29$
F_1	Fischer-Griess	$2^{46} \cdot 3^{20} \cdot 5^9 \cdot 7^6 \cdot 11^2 \cdot 13^3 \cdot 17 \cdot 19 \cdot 23 \cdot 29 \cdot 31 \cdot 41 \cdot 47 \cdot 59 \cdot 71$
F_2	Fischer	$2^{41} \cdot 3^{13} \cdot 5^6 \cdot 7^2 \cdot 11 \cdot 13 \cdot 17 \cdot 19 \cdot 23 \cdot 31 \cdot 47$
F_3	Thompson	$2^{15} \cdot 3^{10} \cdot 5^3 \cdot 7^2 \cdot 13 \cdot 19 \cdot 31$
F_5	Harada-Norton	$2^{14} \cdot 3^6 \cdot 5^6 \cdot 7 \cdot 11 \cdot 19$
HS	Higman-Sims	$2^9 \cdot 3^2 \cdot 5^3 \cdot 7 \cdot 11$
Mc	McLaughlin	$2^7 \cdot 3^6 \cdot 5^3 \cdot 7 \cdot 11$
Suz	Suzuki	$2^{13} \cdot 3^7 \cdot 5^2 \cdot 7 \cdot 11 \cdot 13$
Ly	Lyons	$2^8 \cdot 3^7 \cdot 5^6 \cdot 7 \cdot 11 \cdot 31 \cdot 37 \cdot 67$
He	Held	$2^{10} \cdot 3^3 \cdot 5^2 \cdot 7^3 \cdot 17$
Ru	Rudvalis	$2^{14} \cdot 3^3 \cdot 5^3 \cdot 7 \cdot 13 \cdot 29$
$O'N$	O'Nan	$2^9 \cdot 3^4 \cdot 5 \cdot 7^3 \cdot 11 \cdot 19 \cdot 31$

Table 7.2: Sporadic simple groups

7.3 Affine 2-transitive groups

This table lists the finite 2-transitive groups with elementary abelian socle.

The degree n is a prime power p, and the socle is elementary abelian of order n and regular. The stabiliser $G_0 = H$ of the origin is a subgroup of $\mathrm{GL}(m,p)$. The notation 2^{1+2k} means an extraspecial group of this order (with centre of order 2 and central quotient elementary abelian of order 2^{2k}), and 3 a cyclic group of order 3.

The maximum degree of transitivity is 2 in all cases except the first with $q = 2$ (4-transitive if $d = 2$, 3-transitive if $d > 2$) or $(d,q) = (1,3)$ or $(1,4)$ (3- or 4-transitive respectively), and the case $n = 2^4$, $H = A_7$ (3-transitive).

In the first, fourth, fifth and sixth rows of the table, the conditions are not sufficient for 2-transitivity of G. For example, in the case $n = 3^4$, $P = 2^{1+4} \trianglelefteq H$, we have $H/P \leq S_5$, and G is 2-transitive if and only if H/P contains an element of order 5. For precise conditions in all cases, see Huppert [96], Hering [89].

The fourth and sixth rows include the sharply 2-transitive groups associated with the seven exceptional near-fields (Zassenhaus [192]).

The number of actions is the order of the cohomology group $H^1(G_0, N)$.

Degree	$H = G_0$	Condition	No. of actions
q^d	$\mathrm{SL}(d,q) \leq H \leq \Gamma\mathrm{L}(d,q)$		up to q if q even, $d = 2$
			2 if $(d,q) = (3,2)$
q^{2d}	$\mathrm{Sp}(d,q) \trianglelefteq H$	$d \geq 2$	up to q if q even
q^6	$G_2(q) \trianglelefteq H$	q even	up to q
q	$(2^{1+2} \rtimes 3) = \mathrm{SL}(2,3) \trianglelefteq H$	$q = 5^2, 7^2, 11^2, 23^2$	1
q	$2^{1+4} \trianglelefteq H$	$q = 3^4$	1
q	$\mathrm{SL}(2,5) \trianglelefteq H$	$q = 11^2, 19^2, 29^2, 59^2$	1
2^4	A_6		2
2^4	A_7		1
2^6	$\mathrm{PSU}(3,3)$		2
3^6	$\mathrm{SL}(2,13)$		1

Table 7.3: Affine 2-transitive groups

7.4 Almost simple 2-transitive groups

This table lists the finite 2-transitive groups with simple socle.

Here q is a prime power $q = p^e$; n and d are positive integers, n being the degree. N is the socle of G. The next column gives the index of N in the largest possible G, viz., its normaliser in S_n. Groups with the same socle may have different degrees of transitivity; the least and greatest such degrees are given. Note that the socle is 2-transitive in all cases except one. The final column gives the number of non-isomorphic actions. (For example, for PSL(d, q), $d > 2$, these actions are on the points and hyperplanes of the projective space.)

n	Condition	N	$\max \lvert G/N \rvert$	$\min(t)$	$\max(t)$	No. of actions
n	$n \geq 5$	A_n	2	$n-2$	n	2 if $n=6$ 1 otherwise
$(q^d-1)/(q-1)$	$d \geq 2$ $(d,q) \neq (2,2),(2,3)$	$\mathrm{PSL}(d,q)$	$(d,q-1)e$	3 if $d=2, q$ even 2 otherwise	3 if $d=2$ 2 otherwise	2 if $d>2$ 1 otherwise
$2^{2d-1}+2^{d-1}$	$d \geq 3$	$\mathrm{Sp}(2d,2)$	1	2	2	1
$2^{2d-1}-2^{d-1}$	$d \geq 3$	$\mathrm{Sp}(2d,2)$	1	2	2	1
q^3+1	$q \geq 3$	$\mathrm{PSU}(3,q)$	$(3,q+1)e$	2	2	1
q^2+1	$q=2^{2d+1}>2$	$Sz(q)$	$2d+1$	2	2	1
q^3+1	$q=3^{2d+1}>3$	$R_1(q)$	$2d+1$	2	2	1
11		$\mathrm{PSL}(2,11)$	1	2	2	2
11		M_{11}	1	4	4	1
12		M_{11}	1	3	3	1
12		M_{12}	1	5	5	2
15		A_7	1	2	2	2
22		M_{22}	2	3	3	1
23		M_{23}	1	4	4	1
24		M_{24}	1	5	5	1
28		$\mathrm{PSL}(2,8)$	3	1	2	1
176		HS	1	2	2	2
276		Co_3	1	2	2	1

Table 7.4: Almost simple 2-transitive groups

7.5 Exercises

7.1. Show that $A_2(4) \cong \mathrm{PSL}(3,4)$ and $A_3(2) \cong \mathrm{PSL}(4,2)$ have the same order.

7.2. [CFSG] Show that the only finite simple groups with order coprime to 3 are the Suzuki groups.

7.3. Show that the order of $B_2(q) = \mathrm{PSp}(4,q)$ is a square if $q^2+1 = 2x^2$ for some integer x.

Remark. It is known (using CFSG) that:

(a) the only instances of non-isomorphic finite simple groups of the same order are $|B_n(q)| = |C_n(q)|$ for $n > 2$, q odd, and $|A_2(4)| = |A_3(2)|$;

(b) if S and T are finite simple groups and m and n are positive integers with $|S|^m = |T|^n$, then $m = n$ and $|S| = |T|$.

Part (a) was proved by Artin [5], and updated by various authors as new simple groups were discovered. Part (b) is due to Teague [172], see also [109].

7.4. [CFSG] Use Burnside's Theorem from Section 2.1 and Table 6.4 to determine the transitive permutation groups of prime degree.

7.5. (a) Prove that the length $l(G)$ of a group G is not greater than the number $l_0(G)$ of prime divisors of G (counted according to multiplicity), and that equality holds if and only if it holds for every non-abelian composition factor of G.

(b) [CFSG] Show that a finite simple group with $l(G) = l_0(G)$ is isomorphic to a transitive permutation group of prime degree. Hence show that the only such simple groups are $\mathrm{PSL}(2,2^a)$ (where $p = 2^a+1 \geq 5$ is a Fermat prime), $\mathrm{PSL}(2,7)$, $\mathrm{PSL}(2,11)$, $\mathrm{PSL}(3,3)$, and $\mathrm{PSL}(3,5)$.

7.6. Let G be a group with trivial centre, and suppose that G is generated by two elements g and h. Prove that

$$|C_G(g)| \cdot |C_G(h)| \leq |G|.$$

Hence use CFSG to show that every non-abelian finite simple group G contains an element g with $|C_G(g)| \leq \sqrt{|G|}$.

Bibliography

[1] S. A. Adeleke and Dugald Macpherson, Classification of infinite primitive Jordan permutation groups, *Proc. London Math. Soc.* (3) **72** (1995), 63–123.

[2] S. A. Adeleke and Peter M. Neumann, Relations related to betweenness: their structure and automorphisms, *Memoirs Amer. Math. Soc.* **131**, No. 623 (1998).

[3] Theodore M. Alper, A classification of all order preserving homeomorphisms of the reals that satisfy finite uniqueness, *J. Math. Psychol.* **31** (1987), 135–154.

[4] R. Arratia and S. Tavaré, The cycle structure of random permutations, *Ann. Probability* **20** (1992), 1567–1591.

[5] Emil Artin, The orders of the classical simple groups, *Comm. Pure Appl. Math.* **8** (1955), 455–472.

[6] Michael Aschbacher, On the maximal subgroups of the finite classical groups, *Invent. Math.* **76** (1984), 469-514.

[7] László Babai, On the order of uniprimitive permutation groups, *Ann. Math.* (2) **113** (1981), 553–568.

[8] László Babai, On the length of subgroup chains in the symmetric group, *Comm. Algebra* **14** (1986), 1729–1736.

[9] László Babai and Peter J. Cameron, Automorphisms and enumeration of switching classes of tournaments, to appear.

[10] László Babai, Gene Cooperman, Larry Finkelstein and Ákos Seress, Nearly linear time algorithms for permutation groups with a small base, *Proc. 1991 Internat. Symp. Symbolic and Algebraic Computation*, Bonn, 1991, pp. 200–209.

[11] R. A. Bailey, Restricted randomizaton: a practical example, *J. Amer. Statist. Assoc.* **82** (1987), 712–719.

[12] R. A. Bailey, *Association Schemes and Partial Designs*, in preparation.

[13] Eiichi Bannai and Tatsuro Ito, On finite Moore graphs, *J. Fac. Sci. Univ. Tokyo* **20** (1973), 191–208.

[14] Eiichi Bannai and Tatsuro Ito, *Algebraic Combinatorics I: Association Schemes*, Benjamin, New York, 1984.

[15] Aron Bereczky, Fixed-point-free p-elements in transitive permutation groups, *J. London Math. Soc.* (2) **27** (1995), 447–452.

[16] Claude Berge, Some classes of perfect graphs, Chapter 5 of *Graph Theory and Theoretical Physics* (F. Harary, ed.), Academic Press, New York, 1967.

[17] Meenaxi Bhattacharjee, Dugald Macpherson, Rögnvaldur Möller and Peter M. Neumann, *Notes on Infinite Permutation Groups*, Hindustan Book Agency, 1997.

[18] N. L. Biggs and D. H. Smith, On trivalent graphs, *Bull. London Math. Soc.* **3** (1971), 155–158.

[19] B. J. Birch, R. G. Burns, S. O. Macdonald and P. M. Neumann, On the degrees of permutation groups containing elements separating finite sets, *Bull. Austral. Math. Soc.* **14** (1986), 7–10.

[20] Kenneth D. Blaha, Minimal bases for permutation groups: the greedy approximation, *J. Algorithms* **13** (1992), 297–306.

[21] Andreas Blass and Frank Harary, Properties of almost all graphs and complexes, *J. Graph Theory* **3** (1979), 225–240.

[22] H. F. Blichfeldt, A theorem concerning the invariants of linear homogeneous groups, with some applications to substitution groups, *Trans. Amer. Math. Soc.* **5** (1904), 461–466.

[23] R. C. Bose and D. M. Mesner, On linear associative algebras corresponding to association schemes of partially balanced designs, *Ann. Math. Statist.* **30** (1959), 21–38.

[24] R. C. Bose and K. R. Nair, Partially balanced incomplete block designs, *Sankhyā* **4** (1939), 337–372.

[25] A. E. Brouwer, A. M. Cohen and A. Neumaier, *Distance-Regular Graphs*, Springer, Berlin, 1989.

[26] K. S. Brown, *Buildings*, Springer, New York, 1989.

[27] Francis Buekenhout, Diagrams for geometries and groups, *J. Combinatorial Theory* (A) **27** (1979), 121–151.

[28] William Burnside, *Theory of Groups of Finite Order*, Dover Publications (reprint), New York, 1955.

[29] Peter J. Cameron, Proofs of some theorems of W. A. Manning, *Bull. London Math. Soc.* **1** (1969), 349–352.

[30] Peter J. Cameron, Permutation groups with multiply transitive suborbits, I, *Proc. London Math. Soc.* (3), **25** (1972), 427–440; II, *Bull. London Math. Soc.* **6** (1974), 136–140.

[31] Peter J. Cameron, Transitivity of permutation groups on unordered sets, *Math. Z.* **48** (1976), 127–139.

[32] Peter J. Cameron, Orbits of permutation groups on unordered sets, I, *J. London Math. Soc.* (2) **17** (1978), 410–414; II, *ibid.* (2) **23** (1981), 249–265; III: imprimitive groups, *ibid.* (2) **27** (1983), 229–237; IV: homogeneity and transitivity, *ibid.* (2) **27** (1983), 238–247.

[33] Peter J. Cameron, Finite permutation groups and finite simple groups, *Bull. London Math. Soc.* **13** (1981), 1–22.

[34] Peter J. Cameron, *Oligomorphic Permutation Groups*, Cambridge University Press, Cambridge, 1990.

[35] Peter J. Cameron, Infinite geometric groups of rank 4, *Europ. J. Combinatorics* **13** (1992), 87–88.

[36] Peter J. Cameron, Permutation Groups, Chapter 12 of *Handbook of Combinatorics* (R. L. Graham, M. Grötschel and L. Lovász, eds.), Elsevier, 1995.

[37] Peter J. Cameron, Cofinitary permutation groups, *Bull. London Math. Soc.* **28** (1996), 113–140.

[38] Peter J. Cameron, The random graph, in *The Mathematics of Paul Erdős*, (J. Nešetřil and R. L. Graham, eds.), Springer, Berlin, 1996, pp. 331–351.

[39] Peter J. Cameron, Cycle-closed permutation groups, *J. Algebraic Combinatorics* **5** (1996), 315–322.

[40] Peter J. Cameron, First-order logic, in *Graph Connections: Relationships between graph theory and other areas of mathematics* (L. W. Beineke and R. J. Wilson, eds.), Oxford University Press, Oxford, 1997, pp. 70–85.

[41] Peter J. Cameron and Dmitrii G. Fon-Der-Flaass, Bases for permutation groups and matroids, *Europ. J. Combinatorics* **16** (1995), 537–544.

[42] P. J. Cameron, L. G. Kovács, M. F. Newman and C. E. Praeger, Fixed-point-free permutations in transitive permutation groups of prime power order, *Quart. J. Math. Oxford* (2) **36** (1985), 273–278.

[43] Peter J. Cameron and Jack H. van Lint, *Graphs, Codes, Designs and their Links*, Cambridge University Press, Cambridge, 1991.

[44] Peter J. Cameron and Cleide Martins, A theorem on reconstructing random graphs, *Combinatorics, Probability and Computing* **2** (1993), 1–9.

[45] Peter J. Cameron, Peter M. Neumann and D. N. Teague, On the degrees of primitive permutation groups, *Math. Z.* **180** (1982), 141–149.

[46] P. J. Cameron, C. E. Praeger, J. Saxl and G. M. Seitz, On the Sims conjecture and distance-transitive graphs, *Bull. London Math. Soc.* **15** (1983), 499–506.

[47] Peter J. Cameron, Ronald G. Solomon and Alexandre Turull, Chains of subgroups in symmetric groups, *J. Algebra* **127** (1989), 340–352.

[48] Peter J. Cameron and Csaba Szabó, Independence algebras, in preparation.

[49] Roger W. Carter, *Finite Groups of Lie Type: Conjugacy Classes and Complex Characters*, John Wiley and Sons, Chichester, 1985.

[50] G. L. Cherlin, L. Harrington and A. H. Lachlan, $\aleph_0$-categorical, $\aleph_0$-stable structures, *Ann. Pure Appl. Logic* **28** (1985), 103–135.

[51] J. R. Clay, *Nearrings: Geneses and Applications*, Oxford University Press, Oxford, 1992.

[52] A. M. Cohen and H. Zantema, A computation concerning doubly transitive permutation groups, *J. Reine Angew. Math.* **347** (1984), 196–211.

[53] Paul M. Cohn, *Skew Field Constructions*, London Math. Soc. Lecture Notes **27**, Cambridge University Press, Cambridge, 1977.

[54] J. H. Conway, R. T. Curtis, S. P. Norton, R. A. Parker and R. A. Wilson, *An* **ATLAS** *of Finite Groups*, Oxford University Press, Oxford, 1985.

[55] Bruce Cooperstein, Minimal degree for a permutation representation of a classical group, *Israel J. Math.* **30** (1978), 215–235.

[56] C. W. Curtis, W. M. Kantor and G. M. Seitz, The 2-transitive representations of the finite Chevalley groups, *Trans. Amer. Math. Soc.* **218** (1976), 1–57.

[57] Charles W. Curtis and Irving Reiner, *Representation Theory of Finite Groups and Associative Algebras*, Wiley, New York, 1962.

[58] R. M. Damerell, On Moore graphs, *Proc. Cambridge Philos. Soc.* **74** (1973), 227–236.

[59] Philippe Delsarte, An algebraic approach to the association schemes of coding theory, *Philips Research Reports Suppl.* **10** (1973).

[60] Ph. Delsarte, J.-M. Goethals and J. J. Seidel, Spherical codes and designs, *Geometriae Dedicata* **6** (1977), 363–388.

[61] L. E. Dickson, *Linear Groups, with an Exposition of the Galois Field Theory*, Dover Publications (reprint), New York, 1958.

[62] J. Dieudonné, *La Géometrie des Groupes Classiques*, Springer, Berlin, 1955.

[63] John D. Dixon and Brian Mortimer, The primitive permutation groups of degree less than 1000, *Proc. Cambridge Philos. Soc.* **103** (1988), 213–238.

[64] John D. Dixon and Brian Mortimer, *Permutation Groups*, Springer, 1996.

[65] E. Engeler, Äquivalenzklassen von n-Tupeln, *Z. Math. Logik Grundl. Math.* **5** (1959), 340–345.

[66] Paul Erdős and Alfred Rényi, Asymmetric graphs, *Acta Math. Acad. Sci. Hungar.* **14** (1963), 295–315.

[67] Paul Erdős, Alfred Rényi and Vera T. Sós, On a problem in graph theory, *Studies Math. Hungar.* **1** (1966), 215–235.

[68] David M. Evans, Homogeneous geometries, *Proc. London Math. Soc.* (3) **52** (1986), 305–327.

[69] David M. Evans, Infinite permutation groups and minimal sets, *Quart. J. Math. Oxford* (2) **38** (1987), 461–471.

[70] R. Fagin, Probabilities on finite models, *J. Symbolic Logic* **41** (1976), 50–58.

[71] I. A. Faradžev, M. H. Klin and M. E. Muzichuk, Cellular rings and groups of automorphisms of graphs, in *Investigations in Algebraic Theory of Combinatorial Objects* (I. A. Faradžev, A. A. Ivanov, M. H. Klin and A. J. Woldar, eds.), Kluwer, Dordrecht, 1994, pp. 1–152.

[72] B. Fein, W. M. Kantor and M. Schacher, Relative Brauer groups, II, *J. Reine Angew. Math.* **328** (1981), 39–57.

[73] Walter Feit and Graham Higman, The nonexistence of certain generalized polygons, *J. Algebra* **1** (1964), 434–446.

[74] Walter Feit and John G. Thompson, Solvability of groups of odd order, *Pacific J. Math.* **13** (1963), 775–1029.

[75] Larry Finkelstein and William M. Kantor (eds.), *Groups and Computation*, DIMACS Series in Discrete Mathematics and Theoretical Computer Science **11**, American Mathematical Society, Providence, RI, 1993.

[76] Roland Fraïssé, Sur certains relations qui généralisent l'ordre des nombres rationnels, *C. R. Acad. Sci. Paris* **237** (1953), 540–542.

[77] Roland Fraïssé, *Theory of Relations*, North-Holland, Amsterdam, 1986.

[78] William Fulton, *Young Tableaux*, London Math. Soc. Student Texts **35**, Cambridge University Press, Cambridge, 1997.

[79] M. R. Garey and D. S. Johnson, *Computers and Intractability: An Introduction to the Theory of* NP*-completeness*, W. H. Freeman, San Francisco, 1979.

[80] Y. V. Glebskii, D. I. Kogan, M. I. Liogon'kii, and V. A. Talanov, Range and degree of realizability of formulas in the restricted predicate calculus, *Kibernetika* **2** (1969), 17–28.

[81] L. A. Goldberg and M. R. Jerrum, to appear.

[82] Daniel Gorenstein, *Finite Simple Groups: An Introduction to their Classification*, Plenum Press, New York, 1982.

[83] Daniel Gorenstein, Richard Lyons and Ronald Solomon, *The Classification of Finite Simple Groups*, American Mathematical Society, Providence, RI, 1994.

[84] Ian P. Goulden and David M. Jackson, *Combinatorial Enumeration*, Wiley, New York, 1983.

[85] G. R. Grimmett and D. R. Stirzaker, *Probability and Stochastic Processes*, Oxford University Press, Oxford, 1982.

[86] Mark P. Hale and Ernest E. Shult, Equiangular lines, the graph extension theorem, and transfer in triply transitive groups, *Math. Z.* **135** (1974), 111–123.

[87] Marshall Hall, Jr., On a theorem of Jordan, *Pacific J. Math.* **4** (1954), 219–226.

[88] Frank Harary and Edgar M. Palmer, *Graphical Enumeration*, Academic Press, New York, 1973.

[89] Christoph Hering, Transitive linear groups and linear groups which contain irreducible subgroups of prime order, *Geometriae Dedicata* **2** (1974), 425–460.

[90] D. G. Higman, Intersection matrices for finite permutation groups, *J. Algebra* **6** (1967), 22–42.

[91] D. G. Higman and C. C. Sims, A simple group of order 44,352,000, *Math. Z.* **105** (1968), 110–113.

[92] Graham Higman, Bernhard H. Neumann and Hanna Neumann, Embedding theorems for groups, *J. London Math. Soc.* (1) **24** (1949), 247–254.

[93] Wilfrid Hodges, *Model Theory*, Cambridge University Press, Cambridge, 1993.

[94] Alan J. Hoffman and R. R. Singleton, On Moore graphs of diameters 2 and 3, *IBM J. Research Develop.* **4** (1960), 497–504.

[95] Alexander Hulpke, Konstruktion transitiver Permutationsgruppen, Thesis, Aachen, 1996.

[96] Bertram Huppert, Zweifach transitive, auflösbare Permutationsgruppen, *Math. Z.* **68** (1957), 126–150.

[97] T. Ito and M. Kiyota, Sharp permutation groups, *J. Math. Soc. Japan* **33** (1981), 435–444.

[98] A. A. Ivanov, Distance-transitive representations of the symmetric groups, *J. Combinatorial Theory* (B) **41** (1986), 329–337.

[99] A. A. Ivanov, Distance-transitive graphs and their classification, in *Investigations in the Algebraic Theory of Combinatorial Objects* (I. A. Faradžev, A. A. Ivanov, M. H. Klin and A. J. Woldar, eds.), Kluwer, Dordrecht, 1994, pp. 283–378.

[100] A. A. Ivanov, S. A. Linton, K. Lux, J. Saxl and L. H. Soicher, Distance-transitive represetations of the sporadic groups, *Comm. Algebra* **23** (1995), 3379–3427.

[101] Mark R. Jerrum, A compact representation for permutation groups, *J. Algorithms* **7** (1986), 60–78.

[102] Mark R. Jerrum, Computational Pólya theory, in *Surveys in Combinatorics, 1995* (Peter Rowlinson, ed.), London Math. Soc. Lecture Notes **218**, Cambridge University Press, Cambridge, 1995, pp. 103–118.

[103] Camille Jordan, *Traité des Substitutions et des Équations Algébriques*, Gauthier-Villars, Paris, 1870.

[104] Camille Jordan, Théorèmes sur les groupes primitifs, *J. Math. Pures Appl. (Liouville)* (2) **16** (1871), 383–408.

[105] W. M. Kantor, Jordan groups, *J. Algebra* **12** (1969), 471–493.

[106] W. M. Kantor, Homogeneous designs and geometric lattices, *J. Combinatorial Theory* (A) **8** (1985), 64–77.

[107] W. M. Kantor, Primitive permutation groups of odd degree, and an application to finite projective planes, *J. Algebra* **106** (1987), 15–45.

[108] W. M. Kantor and R. A. Liebler, The rank 3 permutation representations of the finite classical groups, *Trans. Amer. Math. Soc.* **271** (1982), 1–71.

[109] W. Kimmerle, R. Lyons, R. Sandling and D. N. Teague, Composition factors from the group ring and Artin's theorem on orders of simple groups, *Proc. London Math. Soc.* (3) **60** (1990), 89–122.

[110] Peter B. Kleidman and Martin W. Liebeck, *The Subgroup Structure of the Finite Classical Groups*, London Math. Soc. Lecture Notes **129**, Cambridge University Press, Cambridge, 1990.

[111] Serge Lang, *Algebra*, Addison–Wesley, Reading, MA, 1965.

[112] Cristian Lenart and Nigel Ray, Hopf algebras of set systems, *Discrete Math.* **180** (1998), 255–280.

[113] Martin W. Liebeck, On minimal degrees and base sizes of primitive permutation groups, *Archiv der Mathematik* **43** (1984), 11–15.

[114] Martin W. Liebeck, The affine permutation groups of rank 3, *Proc. London Math. Soc.* (3) **54** (1987), 477–516.

[115] Martin W. Liebeck, Cheryl E. Praeger and Jan Saxl, The classification of the maximal subgroups of the finite symmetric and alternating groups, *J. Algebra* **111** (1987), 365–383.

[116] Martin W. Liebeck and Jan Saxl, The primitive permutation groups of odd degree, *J. London Math. Soc.* (2) **31** (1985), 250–264.

[117] Martin W. Liebeck and Jan Saxl, The finite primitive permutation groups of rank 3, *Bull. London Math. Soc.* **18** (1986), 165–172.

[118] Martin W. Liebeck and Jan Saxl, Minimal degrees of primitive permutation groups, with an application to monodromy groups of covers of Riemann surfaces, *Proc. London Math. Soc.* (3) **63** (1991), 266–314.

[119] Martin W. Liebeck and Aner Shalev, Simple groups, permutation groups, and probability, to appear.

[120] D. Livingstone, On a permutation representation of the Janko group, *J. Algebra* **6** (1967), 43–55.

[121] László Lovász, Normal hypergraphs and the perfect graph conjecture, *Discrete Math.* **2** (1972), 253–267.

[122] Heinz Lüneburg, Ein einfacher Beweis für den Satz von Zsigmondy über primitive Primteiler von $a^N - 1$, in *Geometries and Groups*, Lecture Notes in Mathematics **893**, Springer, New York, 1981, pp. 219–222.

[123] Ian G. Macdonald, *Symmetric Functions and Hall Polynomials*, Oxford University Press, Oxford, 1979.

[124] Annabel McIver and Peter M. Neumann, Enumerating finite groups, *Quart. J. Math. Oxford* (2) **38** (1987), 473–488.

[125] Brendan D. McKay, **nauty** user's guide (version 1.5), Technical report TR-CS-90-02, Computer Science Department, Australian National University, Canberra, 1990.

[126] Dugald Macpherson, Infinite distance-transitive graphs of finite valency, *Combinatorica* **2** (1982), 63–69.

[127] Dugald Macpherson, The action of an infinite permutation group on the unordered subsets of a set, *Proc. London Math. Soc.* (3) **46** (1983), 471–486.

[128] Dugald Macpherson, Growth rates in infinite graphs and permutation groups, *Proc. London Math. Soc.* (3) **51** (1985), 285–294.

[129] Dugald Macpherson, Homogeneity in infinite permutation groups, *Period. Math. Hungar.* **17** (1986), 211–233.

[130] Dugald Macpherson, Permutation groups of rapid growth, *J. London Math. Soc.* (2) **35** (1987), 276–286.

[131] Dugald Macpherson, Sharply multiply homogeneous permutation groups, and rational scale types, preprint.

[132] L. Maillet, Sur quelques propriétés des groupes de substitutions d'ordre donnée, *Ann. Fac. Sci. Toulouse*, **8** (1895), 1–22.

[133] W. A. Manning, Simply transitive primitive groups, *Trans. Amer. Math. Soc.* **29** (1927), 815–825.

[134] W. A. Manning, A theorem concerning simply transitive primitive groups, *Bull. Amer. Math. Soc.* **35** (1929), 330–332.

[135] Tracey Maund, D. Phil. thesis, Oxford University, 1989.

[136] H. Nagao, On multiply transitive groups IV, *Osaka J. Math.* **2** (1965), 327–341.

[137] B. H. Neumann, Groups covered by finitely many cosets, *Publ. Math. Debrecen* **3** (1954), 227–242.

[138] Peter M. Neumann, The lawlessness of finitary permutation groups, *Arch. Math.* **26** (1975), 561–566.

[139] Peter M. Neumann, The structure of finitary permutation groups, *Arch. Math.* **27** (1976), 3–17.

[140] Peter M. Neumann, Some primitive permutation groups, *Proc. London Math. Soc.* (3) **50** (1985), 265–281.

[141] A. Yu. Ol'šanskii, An infinite group with subgroups of prime order, *Izv. Akad. Nauk SSSR Ser. Mat.* **44** (1980), 309–321.

[142] A. Yu. Ol'šanskii, *Geometry of Defining Relations in Groups*, Kluwer, Dordrecht, 1991.

[143] Michael E. O'Nan, A characterization of $L_n(q)$ as a permutation group, *Math. Z.* **127** (1972), 301–314.

[144] Michael E. O'Nan, Sharply 2-transitive sets of permutations, in *Proceedings of the Rutgers Group Theory Year* (M. Aschbacher *et al.*, eds.), Cambridge University Press, Cambridge, 1984, pp. 63–67.

[145] James Oxley, *Matroid Theory*, Oxford University Press, Oxford, 1992.

[146] D. S. Passman, *Permutation Groups*, Benjamin, New York, 1968.

[147] Maurice Pouzet, Application de la notion de relation presque-enchaînable au dénombrement des restrictions finies d'une relation, *Z. Math. Logik Grundl. Math.* **27** (1981), 289–332.

[148] Cheryl E. Praeger, Finite transitive permutation groups and finite vertex-transitive graphs, in *Graph Symmetry* (G. Hahn and G. Sabidussi, eds.), Kluwer, Dordrecht, 1997, pp. 277–318.

[149] Cheryl E. Praeger and Jan Saxl, On the orders of primitive permutation groups, *Bull. London Math. Soc.* **12** (1980), 303–307.

[150] C. E. Praeger, J. Saxl and K. Yokoyama, Distance-transitive graphs and finite simple groups, *Proc. London Math. Soc.* (3) **55** (1987), 1–21.

[151] L. Pyber, The orders of doubly transitive permutation groups: elementary estimates, *J. Combinatorial Theory* (A) **62** (1993), 361–366.

[152] L. Pyber, Asymptotic results for permutation groups, in *Groups and Computation*, (Larry Finkelstein and William M. Kantor, eds.), DIMACS Series in Discrete Mathematics and Theoretical Computer Science **11**, American Mathematical Society, Providence, RI, 1993, pp. 197–219.

[153] Richard Rado, Universal graphs and universal functions, *Acta Arith.* **9** (1964), 331–340.

[154] Bruce Reed, A semi-strong perfect graph theorem, *J. Combinatorial Theory* (B) **43** (1987), 223–240.

[155] Mark A. Ronan, *Lectures on Buildings*, Academic Press, Boston, 1989.

[156] G.-C. Rota, On the foundations of combinatorial theory, I: Theory of Möbius functions, *Z. Wahrsch.* **2** (1964), 340–368.

[157] C. Ryll-Nardzewski, On category in power $< \aleph_0$, *Bull. Acad. Pol. Sci. Sér. Math. Astr. Phys.* **7** (1959), 545–548.

[158] Bruce Sagan, *The Symmetric Group: Representations, Combinatorial Algorithms and Symmetric Functions*, Wadsworth, Belmont, 1991.

[159] Rudolf Scharlau, Buildings, in *Handbook of Incidence Geometry* (Francis Buekenhout, ed.), Elsevier, Amsterdam, 1995, pp. 477–645.

[160] L. Schneps (ed.), *The Grothendieck theory of* dessins d'enfants, London Math. Soc. Lecture Notes **200**, Cambridge University Press, Cambridge, 1994.

[161] Martin Schönert *et al.*, GAP — *Groups, Algorithms, and Programming*, Lehrstuhl D für Mathematik, Rheinisch–Westfälische Technische Hochschule, Aachen, Germany, fifth edition, 1995.

[162] Leonard L. Scott, Representations in characteristic p, *Proc. Symp. Pure Math.* **37** (1980), 319–331.

[163] J. J. Seidel, Strongly regular graphs of L_2-type and of triangular type, *Proc. Kon. Nederl. Akad. Wetensch. Ser. A* **70** (= *Indag. Math.* **29**) (1967), 188–196.

[164] J. J. Seidel, A survey of two-graphs, in *Proc. Internat. Coll. Teorie Combinatorie* (Roma 1973), Accad. Naz. Lincei, Roma, 1977, pp. 481–511.

[165] Á. Seress, to appear.

[166] Mark W. Short, *The primitive soluble permutation groups of degree less than 256*, Lecture Notes in Mathematics **1519**, Springer, Berlin, 1992.

[167] Charles C. Sims, Graphs and finite permutation groups, I, *Math. Z.* **95** (1967), 76–86; II, *ibid.* **103** (1968), 276–281.

[168] Derek H. Smith, Primitive and imprimitive graphs, *Quart. J. Math. Oxford* (2) **22** (1971), 551–557.

[169] L.H. Soicher, GRAPE: A system for computing with graphs and groups, in *Groups and Computation* (Larry Finkelstein and William M. Kantor, eds.), DIMACS Series in Discrete Mathematics and Theoretical Computer Science **11**, American Mathematical Society, Providence, RI, 1993, pp. 287–291.

[170] L. Svenonius, $\aleph_0$-categoricity in first-order predicate calculus, *Theoria* **25** (1959), 82–94.

[171] Donald E. Taylor, *The Geometry of the Classical Groups*, Helderman, Berlin, 1992.

[172] David N. Teague, D. Phil. thesis, Oxford University, 1983.

[173] Heiko Theissen, *Eine Methode zur Normalisatorberechnung in Permutationsgruppen mit Anwendungen in der Konstruktion primitiver Gruppen*, Ph.D. Thesis, Aachen, 1997.

[174] Simon R. Thomas, Reducts of the random graph, *J. Symbolic Logic* **56** (1991), 176–181.

[175] Jacques Tits, Généralisation des groupes projectifs basée sur leurs propriétés de transitivité, *Acad. Roy. Belgique Cl. Sci. Mem.* 27 (1952).

[176] Jacques Tits, *Buildings of Spherical Type and Finite BN-Pairs*, Lecture Notes in Mathematics **386**, Springer, Berlin, 1974.

[177] W. T. Tutte, A family of cubical graphs, *Proc. Cambridge Philos. Soc.* **43** (1947), 459–474.

[178] Ascher Wagner, Normal subgroups of triply transitive permutation groups of odd degree, *Math. Z.* **94** (1966), 219–222.

[179] B. Yu. Weisfeiler and A. A. Leman, Reduction of a graph to a canonical form and an algebra which appears in this process, *Scientific-Technological Investigations* Ser. 2, **9** (1968), 12–16 [in Russian].

[180] M. J. Weiss, On simply transitive primitive groups, *Bull. Amer. Math. Soc.* **40** (1934), 401–405.

[181] Dominic J. A. Welsh, *Matroid Theory*, Academic Press, London, 1976.

[182] Helmut Wielandt, Zum Satz von Sylow, I, *Math. Z.* **60** (1954), 407–408; II, *ibid.* **71** (1959), 461–462.

[183] Helmut Wielandt, Primitive Permutationsgruppen vom Grad $2p$, *Math. Z.* **63** (1956), 478–485.

[184] Helmut Wielandt, *Unendliche Permutationsgruppen*, Lecture Notes, Universität Tübingen, 1959.

[185] Helmut Wielandt, Über den Transitivitätsgrad von Permutationsgruppen, *Math. Z.* **74** (1960), 297–298.

[186] Helmut Wielandt, *Finite Permutation Groups*, Academic Press, New York, 1964.

[187] Helmut Wielandt, *Permutation Groups through Invariant Relations and Invariant Functions*, Ohio State University, Columbus, 1969.

[188] Helmut Wielandt, Normalteiler in 3-transitiven Gruppen, *Math. Z.* **136** (1974), 243–244.

[189] Helmut Wielandt, *Mathematische Werke: Mathematical Works*, Volume 1 (Bertram Huppert and Hans Schneider, eds.), Walter de Gruyter, Berlin, 1994.

[190] M. Yoshizawa, On infinite four-transitive permutation groups, *J. London Math. Soc.* (2) **19** (1979), 437–438.

[191] Hans Zassenhaus, Kennzeichnung endlicher linearer Gruppen als Permutationsgruppen, *Abh. Math. Sem. Hamburg* **11** (1936), 17–40.

[192] Hans Zassenhaus, Über endliche Fastkörper, *Abh. Math. Sem. Hamburg* **11** (1936), 187–220.

Index

abelian group, 52
action, 2
Adeleke, S. A., 182
affine group, 9, 16, 28, 104, 181
age, 141
$\aleph_0$-categorical, 139
almost simple, 105
Alper, T. M., 170
alternating group, 15, 27, 109, 165
amalgamation property, 142
antipodal cover, 89
antipodal graph, 89
antipodal quotient, 89
Arratia, R., 177
Artin, É., 198
association scheme, 69
    commutative, 69
    symmetric, 69
ATLAS, 109, 187
automorphism, 34, 43, 64, 138
    graph, 188
    outer, 109

Babai, L., 21, 50, 113, 120
Bailey, R. A., 72, 110
Bannai, E. E., 72, 87, 91
base, 18, 120
basic, 103
basis algebra, 65
basis matrix, 37
Bereczky, A., 176
Berge, C., 155
Bhattacharjee, M., x, 182
Biggs, N. L., 87
bipartite double, 89
bipartite graph, 37, 88
Birch, B. J., 167
Blaha, K., 121
Blass, A., 137
Blichfeldt, H. F., 170
block, 11
Block's Lemma, 43, 92, 93
Bochert, A., 129
boron tree, 150
Bose, R. C., 63, 68, 69, 73
Bose–Mesner algebra, 73
Boston, N., 176
Brauer's Lemma, 93
Brauer, R., 63
Brouwer, A. E., 72, 91
Brown, K. S., 190
de Bruijn, N. G., 37, 143
Buekenhout geometry, 127
building, 190
Burns, R. G., 167
Burnside's Theorem, 36, 101, 110, 198
Burnside, W., 35

Cameron, P. J., x, 21, 32, 50, 74, 110, 111, 118, 123, 124, 156, 168, 184

Cameron's Theorem, 148
Cantor's Theorem, 140
Carter, R. W., 109, 188
cartesian product, 4
Cayley's Theorem, 1, 7
centraliser algebra, 36, 42, 65
Chapman, R. J., 184
character, 40
    irreducible, 40
    multiplicity-free, 46, 67, 120
    permutation, 42
    regular, 45
    sign, 54
character table, 41
Cherlin, G., 182
Cherlin–Mills–Zil'ber Theorem, 182
Chevalley group, 188
chromatic number, 155
class function, 41
classical group, 109, 189
Classification of Finite Simple Groups, 21, 95, 99, 108, 174, 179
Clay, J. R., 17
clique number, 155
cofinitary, 168
Cohen, A. M., 60, 72, 91
coherent configuration, 64
Cohn, P. M., 29
combinatorial geometry, 126
commutative association scheme, 69
Compactness Theorem, 119, 137
congruence, 11
connected, 14
Conway, J. H., 109
core, 5
core-free subgroup, 5
coset space, 4
countably categorical, 139
Curtis, C. W., 40, 110
Curtis, R. T., 109
cycle, 25
cycle decomposition, 25
cycle index, 143, 157
Cycle Index Theorem, 158
cycle structure, 25, 53
cycle-closed, 173

Damerell, R. M., 87
decomposable, 40
degree
    of character, 40
    of matrix representation, 40
    of permutation group, 3
Delsarte, Ph., 69, 91, 92
design, 43
Deza, M., 126
diagonal group, 8, 9, 28, 44, 70, 104
diagonal orbital, 13
diagram, 54
Dickson near-field, 16, 30
Dieudonné, J., 109
directed graph, 13
distance-2 graph, 89
distance-regular graph, 70
distance-transitive graph, 70, 87, 118
Dixon, J. D., x, 16, 17, 113, 165
dual group, 52, 93
dual partition, 53
Dynkin diagram, 188

EIDMA, ix
Engeler, E., 140
enumeration, 37
Erdős, P., 97, 131
Evans, D. M., 179, 183
exceptional group, 109, 190
exceptional near-field, 16

Fagin, R., 137
Fein, B., 174
Feit, W., 29, 74
Feit–Thompson Theorem, 29
fibre bundle, 11, 103
finitary, 165
finitary symmetric group, 165

first-order logic, 135
first-order structure, 136
fixed-point-free, 173
Fon-Der-Flaass, D. G., 124
Fourier transform, 53
Fraïssé class, 142
Fraïssé limit, 142
Fraïssé's Theorem, 142
Fraïssé, R., 141
Frame's condition, 74
Friendship Theorem, 97
Frobenius automorphism, 49
Frobenius group, 168
Frobenius' Theorem, 36, 101
Frobenius, F. G., 36
Frobenius–Schur index, 46, 69
  generalised, 50
Fulton, W., 53
Fundamental Theorem of Abelian Groups, 52

G-space, 2
GAP, 22, 79, 113
Garey, M. R., 39
generalised Frobenius–Schur index, 50
generating function, 144
geometric group, 123, 173
Gilbey, J. D., 59
Glauberman, G., 50
Gleason, A. M., 32
Glebskii, Y. V., 137
global section, 103
Goethals, J.-M., 91
Golay code, 97
Goldberg, L. A., 38
Gorenstein, D., 108
Goulden, I. P., 160
graded algebra, 145
GRAPE, 79
graph, 13
  antipodal, 89
  bipartite, 37, 88
  directed, 13
  distance-2, 89
  distance-regular, 70
  distance-transitive, 70, 87, 118
  Higman–Sims, 83
  Hoffman–Singleton, 78
  hypercubic, 103
  labelled, 39
  Moore, 77, 87
  odd, 96
  orbital, 13
  perfect, 155
  Petersen, 78
  random, 131
  strongly regular, 76
  unlabelled, 39
greedy algorithm, 121
Grimmett, G. R., 38
group
  abelian, 52
  affine, 9, 16, 28, 104, 181
  alternating, 15, 27, 109, 165
  Chevalley, 188
  classical, 109, 189
  diagonal, 8, 9, 28, 44, 70, 104
  dual, 52, 93
  exceptional, 109, 190
  finitary symmetric, 165
  Frobenius, 168
  geometric, 123, 173
  Higman–Sims, 33, 83
  IBIS, 124
  Janko, 33
  Jordan, 179
  Klein, 6
  linear, 22, 28, 101, 109, 181
  Mathieu, 15, 27, 33, 61, 97, 111, 122
  multiply transitive, 110
  of Lie type, 109, 188
  orthogonal, 109

permutation, 2
Ree, 188
sporadic, 109, 192
Steinberg, 188
Suzuki, 188
symmetric, 2, 15, 53, 159
symplectic, 109
twisted, 188
unitary, 81, 109, 168
Zassenhaus, 168

Hadamard product, 91
Hale, M. P., 155
Hamming scheme, 70, 103
Harary, F., 137, 160
Harrington, L., 182
Hering, C., 110
highly set-transitive, 135
highly transitive, 135
Higman's Theorem, 14, 133
Higman, D. G., 14, 33, 63
Higman, G., 8, 74, 84, 154
Higman–Sims graph, 83
Higman–Sims group, 33, 83
HNN-construction, 8
Hodges, W. A., 136
Hoffman, A. J., 77
Hoffman–Singleton graph, 78
holomorph, 9
homogeneous, 133, 141
hook, 55
hook length, 55
Houben, H., x
Howlett, R. B., 110
Hulpke, A., 113
hypercubic graph, 103
hypergraph, 142, 153

IBIS group, 124
imprimitive, 11
incidence structure, 43
indecomposable, 40
induced substructure, 141
integral domain, 146
intersection algebra, 67
intersection matrix, 67
intersection numbers, 64
irreducible character, 40
irreducible representation, 40
Isbell Conjecture, 176
Isbell, J., 176
isomorphic, 3
Ito, T., 72, 87, 91, 171
Ivanov, A. A., x, 120

Jackson, D. M., 160
Janko group, 33
Jerrum's filter, 19
Jerrum, M. R., 19, 38
Johnson scheme, 70
Johnson, D. S., 39
Jordan group, 179
Jordan set, 179
Jordan, C., 99, 113, 165
Jordan–Hölder Theorem, 12
Jordan's Theorem, 100, 127, 179

Kantor, W. M., 110, 168, 174, 179
Kimmerle, W., 198
Kirkman, T. P., 113
Kiyota, M., 171
Kleidman, P. B., 109
Klein group, 6
Klin, M., 176
Kogan, D. I., 137
Kovács, L. G., 184
Krein conditions, 91
Kronecker product, 44, 92
Krull–Schmidt Theorem, 40

labelled, 39
Lachlan, A. H., 182
Lagrange's Theorem, 5
Lang, S., 66
Latin square, 96
left regular representation, 8

Leman, A. A., 63
length, 21
Liebeck, M. W., 108–110, 116, 122, 185
linear group, 22, 28, 101, 109, 181
van Lint, J. H., 74
Linton, S. A., 120
Liogon'kii, M. I., 137
Livingstone, D., 32, 33
Lovász, L., 155
Lucas' Theorem, 7
Lüneburg, H., 175
Lux, K., 120
Lyons, R., 108, 198

McDermott, J. P. J., 149
Macdonald, I. G., 53
Macdonald, S. O., 167
McIver, A., 20
McKay, B. D., 39
Macpherson, H. D., x, 118, 149, 151, 169, 182, 183
MAGMA, 22, 113
Maillet, E., 110, 172
Manning, W. A., 32
Marggraff, B., 180
Markov chain, 38
 aperiodic, 38
 irreducible, 38
Martins, C., 156
Marušič, D., 176
Maschke's Theorem, 40
Mathieu group, 15, 27, 33, 61, 97, 111, 122
Mathieu, É., 113
matrix representation, 40
matroid, 124, 125, 181
Maund, T., 123
membership test, 18
Mesner, D. M., 68, 73
Möbius function, 57
moiety, 182
Möller, R., x, 182
Moore graph, 77, 87
Mortimer, B., x, 16, 17, 113
Müller, J., x, 50, 59, 134
multiplicity-free, 46, 67, 120
multiply transitive, 10, 110

Nagao, H., 31
Nair, K. R., 63, 69
nauty, 39
near-field, 16
Neubüser, J., x
Neumaier, A., 72, 91
Neumann, B. H., 8, 167
Neumann, H., 8
Neumann, P. M., x, 20, 111, 166, 167, 182, 183
Neumann's Lemma, 166, 167
Neunhöffer, M., x, 134
Newman, M. F., 184
Newton's Theorem, 51
Norton, S. P., 109, 187
NP, 39
NP-complete, 39

odd graph, 96
Ol'šanskii, A. Yu., 168
oligomorphic, 135
O'Nan, M. E., 32, 59, 100
O'Nan–Scott Theorem, 105
on-line, 19
orbit, 3
Orbit-Counting Lemma, 37, 47, 176
orbital, 13, 36
 diagonal, 13
 paired, 13
 self-paired, 13
orbital graph, 13
orthogonal group, 109
orthogonality relations, 41, 74
outer automorphism, 109
Oxley, J., 125, 127

P, 39
P and Q matrices, 74
paired orbital, 13
Palmer, E. M., 160
parity, 26
Parker vector, 48
  of character, 50
Parker, R. A., 109, 187
Parker's Lemma, 47, 178
partition, 53
Passman, D. S., x, 17, 129
perfect graph, 155
Perfect Graph Theorem, 155
permutation
  cofinitary, 168
  finitary, 165
permutation character, 42
permutation geometry, 126
permutation group, 2
permutation matrix, 36
Petersen graph, 78
Poincaré series, 146
Poisson distribution, 177
polar space, 190
Pólya, G., 37, 143
Pouzet, M., 153
Praeger, C. E., 108, 113, 118–120, 184
Prime Number Theorem, 111
primitive, 11
product action, 103
projective plane, 34, 79
projective space, 190
Pyber, L., 114

Rado, R., 134, 163
Ramsey's Theorem, 147
random graph, 131
rank, 13
Redfield, J. H., 37, 143
reducible representation, 40
Ree group, 188
Reed, B., 155
regular, 7
regular character, 45
Reiner, I., 40
relational structure, 136
relative Brauer group, 174
Rényi, A., 97, 131
representation
  irreducible, 40
  reducible, 40
Riemann Hypothesis, 111
right regular representation, 7
Robinson–Schensted correspondence, 56
Ronan, M. A., 190
Roney-Dougal, C. M., x
Rota, G.-C., 146
Ryll-Nardzewski, C., 140

Sagan, B., 53, 56
Sandling, R., 198
Saxl, J., 108, 110, 113, 118–120, 185
scale of measurement, 169
Schacher, M., 174
Scharlau, R., 190
Schönert, M., 22
Schreier vector, 17
Schreier, J., 17
Schreier's Conjecture, 109, 130
Schreier's Lemma, 18
Schreier–Sims algorithm, 17
Schur, I., 36
Schur's Lemma, 42
Scott, L. L., 100
Seidel, J. J., 91, 154
Seitz, G. M., 110, 118
self-paired orbital, 13
Semi-Strong Perfect Graph Theorem, 155
semidirect product, 9
semiregular, 7

Seress, Á., 25
set-transitive, 148
Shalev, A., 122
sharp permutation group, 171
sharply multiply transitive, 15, 170
Short, M. W., 113
Shult, E. E., 110, 155
sign, 27
sign character, 54
Sims, C. C., 17, 33, 60, 113
Sims Conjecture, 88, 118
Singleton, R. R., 77
skew field, 29
Smith's program, 88, 118
Smith, D. H., 71, 87, 89
socle, 99
Soicher, L. H., x, 58, 120
Solomon, R., 21, 108
Sós, V. T., 97
sporadic group, 109, 192
stabiliser, 4
Steinberg group, 188
Stirling numbers, 26
Stirzaker, D. R., 38
strong generating set, 18
Strong Perfect Graph Conjecture, 155
strongly closed, 32
strongly connected, 14
strongly regular graph, 76
subcartesian product, 4
subdegree, 13
suborbit, 13
Suzuki group, 188
Svenonius, L., 140
switching, 154
switching class, 154
Sylow's Theorem, 6
symmetric association scheme, 69
symmetric function, 51, 55
symmetric group, 2, 15, 53, 159
symplectic group, 109

tableau, 55
Talonov, D. A., 137
Tarski monster, 168
Tavaré, S., 177
Teague, D. N., 111, 198
Theissen, H., 113
Thomas, S. R., 156
Thompson, J. G., 29, 118
van Tilborg, H., x
Tits, J., 15, 123, 169, 190
topology, 138
transitive, 3
transitive constituent, 4
tree, 29
Turull, A., 21
Tutte, W. T., 87
twisted group, 188
twisted wreath product, 105
two-graph, 154

unitary group, 81, 109, 168
universal, 133
unlabelled, 39

Wagner, A., 32, 100
Weak Perfect Graph Conjecture, 155
weakly closed, 32
Wedderburn's Theorem, 66
Weisfeiler, B. Yu, 63
Weiss, M. J., 33
Welsh, D. J. A., 125, 127
Whitney, H., 125
Wielandt, H., ix, 2, 6, 14, 68, 94, 113, 118, 130, 155, 165, 181
Wilson, R. A., 109, 187
wreath product, 12, 102
  twisted, 105

Yokoyama, K., 119
Yoshizawa, M., 16
Young diagram, 54
Young subgroup, 53
Young tableau, 55

Zantema, H., 60
Zassenhaus group, 168
Zassenhaus, H., 16, 129
zero–one law, 138
Zil'ber, B., 123
Zsigmondy's Theorem, 175

Printed in the United States
By Bookmasters